Problem-Solving and Test-Taking Strategies

JAMESTOWN'S

Number Power

Ellen C. Frechette

JAMESTOWN PUBLISHERS

a division of NTC/CONTEMPORARY PUBLISHING GROUP
Lincolnwood, Illinois USA

ISBN: 0-8092-2279-5

Published by Jamestown Publishers,
a division of NTC/Contemporary Publishing Group, Inc.,
4255 West Touhy Avenue,
Lincolnwood (Chicago), Illinois 60712-1975 U.S.A.

2 3 4 5 6 7 8 9 10 11 12 021 09 08 07 06 05

Table of Contents

To the Student   vi

Pretest   1

BUILDING NUMBER POWER   7

Looking at Word Problems

The Numbers in Your Life   8
Reading Word Problems   10
Multiple-Choice Problems   12
Multistep Word Problems   14
Word Problems with Visuals   16
Using a Calculator   20
What Is Estimation?   22

The Five-Step Process

Solving Word Problems   24
Step 1: Understand the Question   25
Step 2: Find the Information   28
Step 3: Make a Plan   30
Step 4: Solve the Problem   32
Step 5: Check Your Answer   34

Understanding the Question

Read Carefully   36
Some Tricky Questions   41
Working with Set-Up Questions   44
More on Set-Up Problems   47
Mixed Review   52

Finding the Information

Looking at Labels	55
Finding Hidden Information	57
Extra Information	61
Not Enough Information	65
What More Do You Need to Know?	69
Working with Item Sets	71
Making Charts to Solve Item Sets	75
Mixed Review	79

Making a Plan

Choosing the Operation	81
Equations	85
Equations with Two Operations	89
Writing Equations for Word Problems	93
Solving a Word Problem Equation	97
Writing Proportions for Word Problems	101
Solving Proportions	105
When Can You Use a Proportion?	107
Drawing a Picture	110
Mixed Review	115

Solving the Problem

Keeping Organized	117
Estimating the Answer	121
Using Friendly Numbers to Estimate	123
Estimating with Fractions	125
When an Estimate Is the Answer	127
Writing Answers in Set-Up Format	129
Choosing the Correct Expression	131
Comparing and Ordering Numbers	133
Making Good Use of Your Calculator	137
What to Do with Remainders	139
Mixed Review	142

Checking the Answer

Did You Answer the Question?	144
Is the Answer Reasonable?	148

Checking Your Computation 151

Mixed Review 153

Using Charts, Graphs, and Drawings

Reading a Picture 156

Reading Between the Lines on Graphs 160

Reading Graphs and Charts Carefully 163

Mixed Review 167

Working Geometry Word Problems

Finding Information on Drawings 169

Let Formulas Work for You 173

Visualizing Geometry Problems 177

Picturing a Geometry Problem 180

Mixed Review 184

Posttest A 186

Posttest B 192

USING NUMBER POWER 199

Smart Shopping 200

Ordering by Mail 202

Creating a Budget 204

Using a Calculator 206

Using Mental Math 207

Using Estimation 208

Formulas 209

Glossary 210

Index 212

To the Student

Welcome to *Problem-Solving and Test-Taking Strategies*. Whether you are preparing to take a test such as the GED or you are working to improve your math problem-solving skills, you have found the right book.

This book is a little different from other math books you may have used. In this book, you will focus on the **thinking and reasoning** skills that are necessary for success on math tests and in the real world. Yes, you'll have the opportunity to work with decimals, fractions, and percents as well as whole numbers, but the instruction and exercises in this book emphasize the problem-solving process. You will not learn the step-by-step process for multiplying fractions; other Number Power books are designed to help you with this type of math. Instead, *Problem-Solving and Test-Taking Strategies* will teach you to use your common sense and your number sense to find answers to problems.

This book is designed to be used by students of differing mathematical abilities. Each practice problem is accompanied by a symbol that labels its skill level. These symbols will help you select the problems that you have the skill to complete. You will see these symbols with the following skills levels:

WN	whole numbers	**M**	measurements
D	decimals	**GR**	graphs
F	fractions	**GE**	geometry
P	percent		

In the back of this book, you'll find many useful resources to help you as you become a better test taker and problem solver. Important skills such as using a calculator, using mental math, estimating, consulting a glossary, and understanding formulas are taught throughout this book and then summarized in the back. The following icons will alert you to problems in the book where using these skills will be especially helpful.

 calculator icon

 estimation icon

 mental math icon

A chart inside the back cover will help you keep track of your scores.

Pretest

This test will tell you which sections of *Problem-Solving and Test-Taking Strategies* you need to concentrate on. Do every problem that you can. After you check your answers, the chart at the end of the test will guide you to the pages of the book where you need work.

1. If Tammy earns $47.90 in one 4-hour shift, how much does she earn per hour?

   Which of the following is a correct restatement of the question in the problem above?

   a. Find Tammy's pay per day.
   b. Find how much Tammy earns during each shift.
   c. Find Tammy's wages before taxes.
   d. Find how much Tammy earns in one year.
   e. Find how much Tammy is paid for one hour.

2. Carmine drives 6 miles to work each day and the same distance home again. If he works 4 days this week, how many miles did he drive in all for work?

   a. 6   d. 24
   b. 10   e. 48
   c. 12

3. According to the chart at the right, how many pounds can a 3-cubic-foot crate hold?

   a. 60   d. 180
   b. 100   e. 200
   c. 140

   ## Packing Guidelines

Crate	Weight
1 cu ft	60 lb
2 cu ft	100 lb
3 cu ft	140 lb
4 cu ft	200 lb

4. Kay planted 6 rows of lettuce with 6 plants in each row. Each row was 6 feet long, and it took her 6 hours to do all the planting. How many lettuce plants did she plant in all?

   The following information is **not** needed to solve the problem:

   a. 6 feet, 6 hours
   b. 6 plants
   c. 6 feet, 6 plants
   d. 6 rows, 6 plants
   e. 6 rows

5. Max spent 18 hours assembling 3 picnic tables. On average, how many hours did each table take to assemble?

   Which of the following operations should you use to solve the problem above?

   a. Multiply 18 by 3.
   b. Divide 18 by 3.
   c. Subtract 3 from 18.
   d. Add 3 and 18.
   e. Divide 3 by 18.

6. The Halt Hunger Telethon raised $300,000 from about 10,000 callers. Approximately how much money did each caller pledge?

   Without doing any calculations, choose the most sensible answer from below.

   a. $10
   b. $30
   c. $3,000
   d. $300,000
   e. $3,000,000

7. According to the chart at the right, how many *yards* of twine will the packer need to wrap three #2 cartons?

   a. 3
   b. 4
   c. 5
   d. 6
   e. 12

Twine Lengths	
Carton	Number of Feet
#1	2
#2	4
#3	6
#4	8

8. How many *cups* of liquid are held in the containers shown at the right?

   a. $2\frac{1}{2}$
   b. 6
   c. 8
   d. 10
   e. not enough information given

   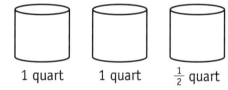

   1 quart   1 quart $\frac{1}{2}$ quart

9. Lisa paid $55 for a dress at Discount Dan's and $90 for a dress at Mrs. Mooney's Shop. Which of the following expressions shows how much more Lisa paid for the more expensive dress?

   a. $90 + $55
   b. $55 − $90
   c. $90 − $55
   d. $90 ÷ $55
   e. $90 × $55

10. A mother baked 70 brownies and gave 12 of them to each of her three neighbors. Which of the following expressions shows how many brownies she had left?

   a. $70 - (3 \times 12)$
   b. $70 - 12$
   c. $(70 - 12) \times 3$
   d. $(70 - 12) \div 3$
   e. $70 - 12 - 3$

11. According to the chart at the right, which of the following expressions shows the average weight in pounds of the Sammarco children at birth?

   a. $6 + 7.5 + 7.5 + 8$

   b. $\dfrac{6 + 7.5 + 7.5 + 8}{3}$

   c. $\dfrac{6 + 7.5 + 7.5 + 8}{4}$

   d. $3 \times (6 + 7.5 + 7.5 + 8)$

   e. $4 \times (6 + 7.5 + 7.5 + 8)$

Sammarco Family	
Jennifer	— 6 lb
Meghan	— 7.5 lb
Jonathan	— 7.5 lb
Matthew	— 8 lb

12. There are twenty-eight people on Floyd's shift at the factory. Half of them are women. How many women are on Floyd's shift?

   a. 2
   b. 14
   c. 28
   d. 56
   e. not enough information given

13. A manufacturing plant turns 9,400 circuit boards per shift. If there are 10 assembly lines running on each shift, with 5 employees per line, how many boards does each assembly line turn out per shift?

   a. 188
   b. 940
   c. 1,800
   d. 47,000
   e. 94,000

Problems 14 and 15 refer to the graph at the right.

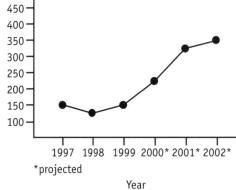

Meals on Wheels: Marshtown
Program Participants
1997–2002*

14. According to the graph, about how many more people in Marshtown used the Meals on Wheels in 1999 than in 1998?

   a. 25
   b. 120
   c. 125
   d. 150
   e. not enough information given

15. According to the graph, how many people over the age of 65 are expected to use the Meals on Wheels program in 2001?

   a. 350
   b. 325
   c. 300
   d. 225
   e. not enough information given

16. Mary and Lewis were married in 1988. Three years later they had their first baby. What more do you need to know to find out how old Mary was when she had her first child?

    a. Mary's age when she got married
    b. Lewis's present age
    c. the birth date of Mary's child
    d. the present age of Mary's child
    e. the year the baby was born

Problems 17–19 refer to the following information.

   To landscape their yard, Maureen and David bought 4 cubic yards of loam at $100.00 per cubic yard. From the same landscaping company, they purchased 4 evergreen bushes at $15.99 each and a dozen smaller shrubs at $7.98 each.
   To have the landscaping company spread the loam and plant the shrubs and bushes, it would cost Maureen and David $35.00 hour for labor.

17. How much money did Maureen and David spend on the bushes and shrubs?

    a. $63.96   d. $159.72
    b. $95.76   e. not enough information given
    c. $109.72

18. The landscaping company estimates that it would take 22 hours to spread the loam and plant the shrubs. How much would the company charge for the labor?

    a. $35.00   d. $879.72
    b. $720.00   e. $929.72
    c. $770.00

19. Which expression shows how much the landscaping company would charge to do the whole job, including materials and 22 hours of labor?

    a. $22 \times \$35.00$
    b. $4(\$100.00 + \$15.99) + (22 \times \$35.00)$
    c. $(4 + \$100.00) + (4 + \$15.99) + (12 + \$7.98) + (22 + \$35.00)$
    d. $(4 \times \$100.00) + (4 \times \$15.99) + (12 \times \$7.98) + (22 \times \$35.00)$
    e. $(4 \times \$100.00) + (4 \times \$15.99) + \$7.98 + (22 \times \$35.00)$

20. Peter bought a total of 13 baseball game tickets and gave some to Sondra. He found that he had 9 left. Which of the following equations will tell you how many tickets Peter gave Sondra?

    a. $13 + 9 = x$
    b. $9 + 13 = x$
    c. $9 - x = 13$
    d. $x + 13 = 9$
    e. $13 - x = 9$

21. To make the figure shown at the right, Su Lin needs 5 tablespoons of rubber cement. How many tablespoons of rubber cement would she need for a similar figure that measures 84 inches around?

    a. 5
    b. 15
    c. 56
    d. 112
    e. 127

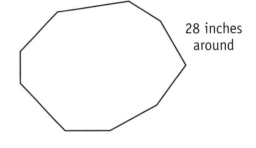

28 inches around

22. Approximately how many loads of laundry do the Mosers do each year if they wash 5 loads per week?

    a. 35
    b. 50
    c. 250
    d. 500
    e. 2,500

23. The chart at the right shows the amount of time different participants took to complete a puzzle for a scientific research project. Which of the following shows the correct order from *shortest* to *longest* time?

    a. Aaron, Patrick, Esther, Nora, Charlie
    b. Charlie, Nora, Esther, Patrick, Aaron
    c. Aaron, Esther, Nora, Patrick, Charlie
    d. Esther, Nora, Patrick, Charlie, Aaron
    e. not enough information given

Experiment 12a7

Participant	Time
Esther	35 min
Aaron	28 min
Patrick	$\frac{1}{2}$ hr
Nora	36 min
Charlie	1 hr

24. According to the graph at the right, how many more Norton residents lived in subsidized housing in 1999 than 1998?

    a. 1
    b. 4
    c. 7
    d. 1,000
    e. 4,000

Norton Residents Living in Subsidized Housing/Rent-Controlled Housing (1998–2000)

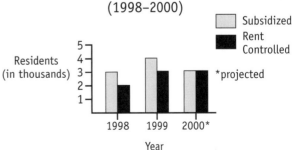

□ Subsidized
■ Rent Controlled

*projected

25. Which of the following formulas can be used to find the number of square units in the figure at the right?

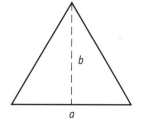

a. $P = a + b + c$
b. $P = \frac{1}{2}(a + b + c)$
c. $A = a \times b \times c$
d. $\frac{1}{2}(a \times b \times c)$
e. $A = \frac{1}{2}ab$

Pretest Chart

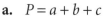

If you miss more than one problem in any section of this test, you should complete the lessons on the practice pages indicated on this chart. If you miss only one problem in a section of this test, you may not need further study in that chapter. However, to effectively master problem-solving and test-taking strategies, we recommend that you work through the entire book. As you do, you should focus on the skills for the items that you missed.

PROBLEM NUMBERS	SKILL AREA	PRACTICE PAGES
2	multistep problems	14–15
1, 7, 8	understanding the question	25–27, 36–51
9, 10, 11, 19	set-up questions	44–51
12	finding hidden information	57–60
4, 13	extra information	61–64
15, 16	not enough information	65–68
17, 18	working with item sets	71–78
5	making a plan	30–31, 81–114
20	writing equations for word problems	93–100
21	writing proportions for word problems	101–106
14, 22	using estimation	121–128
23	comparing and ordering	133–136
6	reasonable answers	148–150
3, 24	using charts, graphs, and drawings	156–166
25	using formulas	173–176

Building Number Power

LOOKING AT WORD PROBLEMS

The Numbers in Your Life

| 475 | Out of a total of 475 people who signed a petition, 390 |
| − 390 | were women. How many men signed the petition? |

How are these two math problems the same?
How are they different?

The big difference is that the first problem does not use any words. It simply gives you two numbers and tells you to subtract.

The second problem is a **word problem.** It describes a situation and asks a question. It gives you two numbers, just as the first problem does. However, in a word problem *you* must figure out whether to add, subtract, multiply, or divide.

For both of these problems you would subtract to get a correct answer. But with a word problem you need to take other steps *before* you subtract. Figuring out these steps is what makes word problems challenging.

Imagine driving along a highway, trying to figure out how many miles you have driven. You know the total distance to Mayfield, your destination, is 15 miles, and you see a sign that says, "Mayfield: 6 miles."

The sign is giving you the information you need to find out how far you have driven. However, the sign does *not* tell you to subtract those 6 miles from the total of 15 miles. You must decide for yourself what to do with the numbers.

The number problems you come across in your everyday life are usually similar to word problems. You solve math problems with a little bit of common sense and a little bit of "number sense."

You already have acquired a lot of common sense. How do you get "number sense"? By working with numbers. By trying new things with numbers. By seeing how numbers play a part in your life.

Think about the numbers you deal with each day and jot them down.

Your age: _____   The year you were born: _____

Your height: _____   A friend's height: _____

Your weight last year: _____   Your weight this year: _____

The time you wake up: _____   The time you go to bed: _____

Your salary: _____   Your rent: _____

Amount of money
you have right now: _____   Amount of money you spend
on food each week: _____

Cost of a loaf of bread: _____   Cost of a gallon of milk: _____

Use the numbers above to write five word problems. Make them as easy or as hard as you like—because you won't have to solve them! An example is given below to help you get started.

EXAMPLE   Each morning I wake up at 7:00 A.M., and I go to bed at 10:00 P.M. each night. How many hours am I awake each day?

1. _____

_____

2. _____

_____

3. _____

_____

4. _____

_____

5. _____

_____

Reading Word Problems

When people read word problems, they think, "Hey, this is a math problem. I'd better concentrate on the numbers." When you read a word problem for the first time, concentrate on the **story,** not the numbers. If you first picture what is going on in the problem, you can more easily and accurately figure out what to do with the numbers.

Look at the word problem below. Then, **without using any numbers,** answer the questions that follow.

A welder needed a pipe 30 inches long to fit a newly installed sink. He cut the piece he needed from a $3\frac{1}{2}$-foot pipe. How many inches long was the remaining piece of pipe?

What did the welder need?
Where did he get it?
What is the question?

Now use your answers to reword the problem **using no numbers.** Replace the numbers with general words like *some, several, a few, a certain number of,* and *more.* You'll probably come up with something like this:

A welder needed *some* pipe. He cut it from *another* pipe. *How much pipe was left over?*

Can you picture what is happening in this problem? Before you focus on numbers and feet and inches, ask yourself, "What is going on in the problem?"

Practice reading the word problems on the next page without concentrating on the numbers. On a separate sheet of paper or out loud with a partner, reword each problem without using any numbers. Replace the actual numbers with words like *some, several, a few,* and *more.* Try to picture what is happening in each story. Do not solve the problems.

EXAMPLE   Paul offers Naomi $0.50 for every mistake she can find in his essay. After reading it carefully, Naomi finds 13 mistakes. How much money does Paul owe Naomi?

Paul offers Naomi some money for every mistake she finds. She finds many mistakes. How much does Paul owe Naomi?

WN 1. The Crate Company hired 17 new men to work in its warehouse division. It also hired 25 women for the inventory department. How many people were hired altogether?

WN 2. A group of 72 committee members wanted to break down into smaller work groups. If they wanted a total of 8 groups, how many committee members made up a group?

WN 3. Over a 12-hour period, a nurse at the Alcohol Abuse Hot Line received 35 calls. If she spent an average of 7 minutes on each call, approximately how many hours in all did she spend on the telephone?

WN 4. For the past 7 games, batter José Alvarez has had 3 hits, 5 hits, 7 hits, 0 hits, 2 hits, 1 hit, and 0 hits. What is his average number of hits for the last 7 games.

D 5. A tailor uses up twelve dozen spools of thread each week. If each spool costs $1.19, how much does the tailor spend in thread each week?

D 6. Seventeen mail carriers each worked an 11-hour day on Tuesday, due to a holiday backlog of mail. If each mail carrier is paid $12.90 per hour, what was the total paid in wages for these seventeen carriers on Tuesday?

D 7. When Todd works overtime at the plant, he earns $12.50 per hour. His regular hourly wage is $9.90. How much more does Todd earn per hour when he works overtime?

F 8. A woman started a trip driving 55 mph, and she drove for 3 hours at this rate. Toward the end of her trip she picked up speed and drove 62 mph for an hour and a half. How many miles did the woman drive altogether on this trip?

Multiple-Choice Problems

At Chase's Market, celery is sold for $0.89 per pound. A larger supermarket in the same neighborhood sells celery for $0.78 per pound. If Robert bought 3 pounds of celery at the supermarket, how much money did he save by not buying it at Chase's?

a. $0.11   **d.** $2.34

b. $0.27   **e.** $2.67

c. $0.33

Why is it that some word problems include a choice of answers? Do you think answer choices make the problem easier?

Word problems like the one above are **multiple choice.** (The correct answer is **c. $0.33.**) With these problems you are given a list of answers and you must choose the correct one. A list of answers does not necessarily make your job easier. You still must **solve** the word problem as you would if no answer choices were given.

In the problem above, four steps can be taken to answer the question.

STEP 1	Find out how much the celery would cost at Chase's.	*$0.89 × 3 = $2.67*
STEP 2	Find out how much the celery would cost at the larger supermarket.	*$0.78 × 3 = $2.34*
STEP 3	Subtract the supermarket price from Chase's price.	*$2.67 − $2.34 = $0.33*
STEP 4	Find the matching answer.	*c. $0.33*

Did You Know . . . ?

- Tests have a multiple-choice format because they are easier to score that way.

- On multiple-choice tests each *wrong* answer listed is carefully chosen. For example, if you added when you should have subtracted, you might find your *incorrect* answer listed as one of the choices!

- Simply guessing an answer is not a good strategy, but there are ways you can eliminate one or two of the answer choices. You'll learn more about these strategies later in this book.

For most of your work in this book, you should solve the word problems without paying too much attention to the answer choices. A list of answer choices is **not** an invitation to guess blindly.

First read the following word problems. Then imagine you are writing a multiple-choice test and list three possible answer choices for each problem. Make sure one of the answers is the correct one!

EXAMPLE   Bill weighs 196 pounds right now. He would like to lose 16 pounds. What is Bill's desired weight?

a. _____212_____   b. _____180_____   c. _____3,136_____

↑   ↑   ↑

(add)   (subtract)   (multiply)

WN 1. Joanna put $127 worth of clothing on layaway with a $10 down payment. The next week she came into the store and put down another $10. Last Thursday she put down $25 more. How much more money does she owe for the clothing?

a. _____   b. _____   c. _____

WN 2. On Halloween, Mr. Dody put together all the candy bars collected by each of his three children. Then he gave each child one candy bar per day for as long as the candy lasted. Patrick collected 20 bars, Peter collected 14 bars, and Jack collected 11 bars. On average, how many days did the candy last?

a. _____   b. _____   c. _____

F 3. The two pitchers that Mrs Sylvia is using hold $4\frac{1}{2}$ quarts and $1\frac{1}{2}$ quarts. How many more quarts does the larger pitcher hold?

a. _____   b. _____   c. _____

P 4. Nancy left $500 in a savings account for one year. The bank pays 7% interest annually. What total amount of money did Nancy have at the end of the year?

a. _____   b. _____   c. _____

GE 5. Roberto enclosed the rectangular garden at the right with wire mesh. How many feet of mesh did he need?

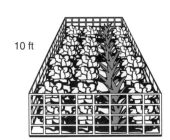

10 ft

5 ft

a. _____   b. _____   c. _____

Multistep Word Problems

On Thursday, Tom collected 55 cans; on Friday, he collected 65 cans. How many cans did he collect in all?

On Thursday, Tom collected 55 cans; on Friday, he collected 65 cans. He took them to the recycling center, where he was paid $0.05 per can. How much did Tom earn for the two days?

How are these two problems different? What are the steps in solving each?

The first step in both problems is the same.

Problem 1	Problem 2
55 cans	55 cans
+ 65 cans	+ 65 cans
120 cans	120 cans

In fact, **120 cans** is the answer to the first problem. Your only step was to add 55 and 65. However, for the second problem you need to perform another step.

The second problem is a **multistep** problem. This means that to get the correct answer you must perform more than one operation. Many word problems are multistep problems. You'll get plenty of practice with them in this book.

Problem 2
120 cans
× 0.05 per can
$6.00 total

Solve the following word problems. Show your work for each step and circle your final answer.

EXAMPLE   Andrea bought 2 bunches of flowers for $2.99 each. She also bought a vase for $5.00. How much did Andrea spend in all?

STEP 1
$2.99
× 2 bunches
$5.98

STEP 2
$5.98 flowers
+ 5.00 vase
$10.98 in all

WN 1. During a flu epidemic at the plant, 36 of the company's 120 workers were out sick one day. To increase production, the company hired 20 temporary workers for the day. How many people were working at the plant that day?

Step 1   **Step 2**

D 2. A customer purchases a box of toothpicks for $0.72 plus a tax of $0.04. She pays with a $10 bill. How much change should she receive from the cashier?

Step 1   **Step 2**

D 3. To save money Louise decides to walk instead of paying the $0.75 bus fare each way to and from work. After 14 workdays, how much money will she have saved?

Step 1   **Step 2**

F 4. Hal works $8\frac{1}{2}$ hours per day, 5 days per week. How many hours does he work in a 4-week month?

Step 1   **Step 2**

F 5. Sarah bought $1\frac{1}{2}$ pounds of apples and $3\frac{1}{2}$ pounds of oranges. How much did she pay?

Step 1   **Step 2**   **Step 3**

APPLES
$1.38/lb

ORANGES
$1.34/lb

GE 6. The Duffys' garden measures 20 feet by 40 feet. They use 2 scoops of fertilizer for every 100 square feet of garden. How many scoops do they need for the whole garden?

Step 1   **Step 2**   **Step 3**

Word Problems with Visuals

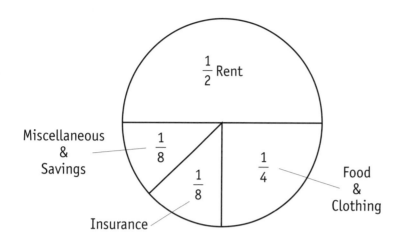

The Simsar family's total monthly income is $3,000. How much do they budget for food and clothing each month?

a. $3,000 **d.** $750

b. $2,000 **e.** $500

c. $1,500

What information would you use to answer the word problem above?

Many word problems require you to find and use information on a graph, chart, map, or picture in order to find a solution.

To solve this problem you need to take information **from the graph** and **from the problem** itself.

$$\frac{1}{4} \qquad \times \qquad \$3{,}000 \qquad = \quad \$750$$

$\downarrow$	$\downarrow$
on the graph	in the problem
as	as
Food & Clothing	**monthly income**

The correct answer choice is **d. $750**.

Let's look at another example of a word problem that includes a **visual.**

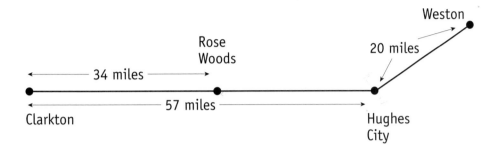

On a trip from Clarkton to Hughes City, Henry stopped in Rose Woods for a bite to eat. How many more miles did he have to drive after the stop?

As you can see, this word problem is accompanied by a map. You must look at the map and the mileage given to answer the question.

What information do you need from the map to answer the question?
What is the correct answer?

57 miles	–	34 miles	=	**23 miles**
↓		↓		↓
total distance Clarkton to Hughes City		Clarkton to Rose Woods		remaining distance to Hughes City

The distance from Hughes City to Weston, 20 miles, is not needed to answer the problem. Many visuals include more information than is necessary. You must be careful to choose only the numbers that you need to solve a problem.

Solve the following word problems based on the chart below.

Year	North America	South America	Europe
Estimated Population by Region (in millions)			
1900	106	40	400
1950	221	100	392
1980	252	242	484
1998	301	508	508

WN 1. How much greater was the population in Europe in 1998 than in 1980?

    a. 24   **d.** 24,000,000

    b. 240   **e.** 37,000,000

    c. 992

WN 2. In 1998 how much smaller was the population of North America than the population of Europe?

    a. 207   **d.** 207,000

    b. 2,070   **e.** 207,000,000

    c. 20,700

F 3. In 1900 the population of North America was approximately what fraction of the population of Europe?

    a. $\frac{3}{4}$ **d.** $\frac{1}{4}$

    b. $\frac{1}{2}$ **e.** $\frac{1}{8}$

    c. $\frac{3}{8}$

P 4. By what percent did the population of South America grow between 1900 and 1950?

    a. 7.5%   **d.** 100%

    b. $33\frac{1}{3}$%   **e.** 150%

    c. 50%

Use the graph below to answer the questions that follow.

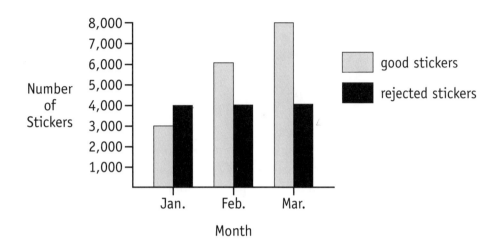

WN 5. In January how many more rejected stickers than good stickers were produced?

   a. 4,000   **d.** 1,000

   b. 3,000   **e.** 0

   c. 2,000

WN 6. How many good stickers were produced in all during the first quarter of 1999?

   a. 17,000   **d.** 4,000

   b. 12,000   **e.** 2,000

   c. 5,000

P 7. By what percent did the number of good stickers increase from January to February?

   a. 3,000%   **d.** 50%

   b. 300%   **e.** 5%

   c. 100%

Using a Calculator

Maggie read the first 88 pages of a chapter book. Jack read the next 95 pages, and Anna read the last 127 pages. How many pages did the children read in all?

a. 222   **b.** 237   **c.** 310   **d.** 315   **e.** 410

> **Would a calculator help you solve this word problem more quickly?**
> **Would your answer be more accurate?**

On many tests, and certainly in everyday life, use of a calculator is allowed and even encouraged. A calculator cannot take the place of thinking through how to solve a problem. However, once you have a plan for solving a problem, a calculator can help you work with a challenging computation.

Let's take a look at how a basic calculator works.

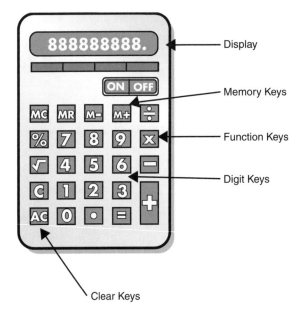

All calculators have **digit keys** (numbers) and **function keys** (basic operations such as addition, multiplication, subtraction, and division). Also, a calculator has a **display** that shows values in whole numbers and decimals.

When you turn on the calculator, the number **0** appears on the display. Each time you enter a digit, the number will appear on the display.

Here are the steps to use a calculator to solve the world problem on the previous page. You have probably already decided to *add* the three numbers together to find a sum.

127 + 88 + 95 =

Press these keys:	**Calculator displays:**
1 2 7	127.
+	127.
8 8	88.
+	215.
9 5	95.
=	310.

The children read a total of **310 pages.**

You can also use a calculator to solve decimal problems. Look at this example and think about what keys you would press on the calculator.

350.5 ÷ 0.25 =

Press these keys:	**Calculator displays:**
3 5 0 . 5	350.5
÷	350.5
. 2 5	0.25
=	1402.

 Use a calculator to solve the following problems.

1. $3,017 - 299 =$

2. $4,032 / 96 =$

3. $286 + 138 + 37 =$

4. $716 \times 41 =$

5. Larry bought a CD for $12.99, a CD rack for $22.95, and a package of blank cassettes for $5.99. How much money did he spend in all?

6. A carpenter cut a 15-meter piece of plywood into lengths of 1.2 meters. How many 1.2-meter pieces did he end up with?

What Is Estimation?

A customer paid for 3 bags of potato chips with a $20 bill. Each bag cost $0.89. How much change should the customer receive?

a. $22.67   **b.** $19.11   **c.** $17.33   **d.** $2.67   **e.** $1.50

Can you take a good guess on the answer to this problem? Do you really need to multiply 0.89 by 3 to find the correct answer?

Estimation is a skill that you probably already use in your everyday life. It is also a skill that is very useful when you take a test. When you estimate, you round numbers to make them easier to work with. You then get an answer that is "close" to the actual answer.

For example, in the problem above, what would happen if you said that $0.89 is about $1.00? You could then say that the chips cost about $3.00, right?

$20.00 – $3.00 = $17.00

You have figured out that the customer should get about $17.00 in change from his purchase. Which answer choice is closest to $17.00?

You are correct if you chose **c.** $17.33. Now check by doing the computation.

$0.89 $20.00
×   3   – 2.67
―――   ―――
$2.67   **$17.33**

When you are doing any kind of math problem, look for numbers that can be changed slightly to make them easier to work with. Just remember that your answer will be only an estimate, not an exact answer. But in many cases, an estimate is all you will really need.

Practice estimating by finding easy numbers that the following numbers are close to.

EXAMPLE   38 is close to 40.

1. 88 is close to ___.

2. 1,001 is close to ___.

3. 82 is close to ___.

4. 58 is close to ___.

5. 196 is close to ___.

6. 149 is close to ___.

7. 3,899 is close to ___.

8. 203 is close to ___.

Find an estimate for each problem below. Remember to use numbers that are easy to work with.

EXAMPLE   A 185-foot antenna was placed atop a building that is 804 feet high. How many feet off the ground is the tip of the antenna?

185 is about 200. 804 is about 800.
200 + 800 = **1,000 feet high**

9. Lyle drove 21 miles on Monday, 48 miles on Tuesday, 97 miles on Wednesday, and 101 miles on Thursday. Approximately how many miles did he drive in all?

Estimate: _____

10. A 79-inch rubber tube is divided into 4 equal lengths. About how long is each piece of rubber?

Estimate: _____

11. Susan bought 19 jars of baby food for her daughter Ashley. Each jar cost $0.98. Approximately how much money did Susan spend on baby food?

Estimate: _____

12. It took Bev 34 minutes to drive to school to pick up Doug. She then drove 58 minutes to get to Hannah's guitar lesson, and another 29 minutes home. About how many *hours* was Bev in the car?

Estimate: _____

THE FIVE-STEP PROCESS

Solving Word Problems

When you use a **process** to get something done, you take a series of steps, in a certain order, to reach your goal. For example, you must correctly follow a process to bake a cake, get a new job, or change the oil in your car.

> **What Would Happen If . . .**
>
> • you put a cake mix in the oven *before* you added water and poured it into a pan?
>
> • you skipped the interview and application and just showed up for a new job?
>
> • you poured new oil into your car without emptying the old oil first?

Solving word problems is a **process** too. Nobody can just jump in and solve a word problem in one step. Instead, you must follow a series of steps, one at a time, before you can arrive at the correct solution.

For the rest of this book you will be working with different ways to solve word problems. One of the main methods is the five-step problem-solving process, illustrated below.

The Five-Step Problem-Solving Process

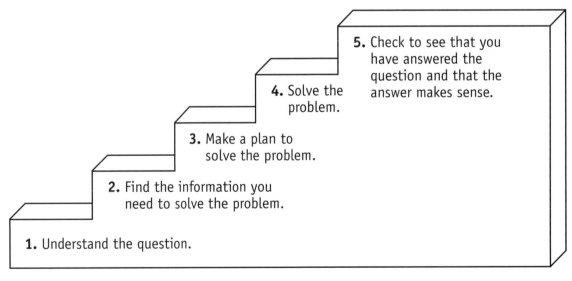

5. Check to see that you have answered the question and that the answer makes sense.

4. Solve the problem.

3. Make a plan to solve the problem.

2. Find the information you need to solve the problem.

1. Understand the question.

Step 1: Understand the Question

It took Nita's secretary 35 minutes to type a report. If the secretary can type 65 words per minute, how many words were in Nita's report?

What does the problem ask you to find?

No matter how many sentences or numbers a word problem has, it *always* asks you to find something. This problem asks you to *find the number of words in the report.*

Sometimes a problem will *ask* a question, as in the problem above. Other times a problem will *tell* you what to find.

The bell in a church tower rings 4 times every hour from 8:00 in the morning until 8:00 at night. It then rings once per hour for the remaining 12 hours. Find how many times the bell rings in a 24-hour period.

Although there is no question mark in this problem, you *are* asked to find something. To do the problem you must *find how many times the bell rings.*

Before you can solve a problem, you must first make sure you understand the question. Otherwise, you will start your work like a person who starts a trip with no idea of the destination.

The chapter called Understanding the Question gives you strategies and practice to help you understand the question.

Read each word problem below and write down, in your own words, what you are being asked to find. It may help to underline key words as you read. Do not copy the questions and do not solve the problems.

EXAMPLE   Troy paid $0.85 for each of 6 brownies at a bake sale. How much did Troy spend altogether on the brownies?

FIND: *How much Troy spent on brownies*

WN 1. Ten Boy Scouts participated in a fund-raising drive. What amount of money did the Boy Scouts collect if each boy turned in $15?

FIND: _____

WN 2. Barbara hired 3 assistants to correct some exams. There are 63 exams to correct in all. If Barbara divides the exams equally among the assistants, how many exams will each assistant correct?

FIND: _____

D 3. When she worked last Saturday, Marlene earned $32.00 in tips during her 4-hour shift. If her hourly wage is $3.35, how much did she earn during that shift?

FIND: _____

P 4. A hairdresser in Vera's Salon gave 3 haircuts at $24.00 each and 4 permanents at $55.00 each. If the salon takes 60 percent of the hairdresser's payments, how much money will it collect from the hairdresser's work?

FIND: _____

F 5. Bananas cost $0.59 per pound. Maria purchased $1\frac{1}{2}$ pounds of bananas. How much did she spend?

FIND: _____

Finish writing each word problem below by adding a question. There is no single correct way to complete the problems, so use your imagination.

EXAMPLE   Alex Moser weighs 28 pounds. His sister Juliette weighs 38 pounds.

How much more does Juliette weigh than Alex?

F   **6.** Lin Chung spent $1\frac{1}{2}$ hours studying math, $\frac{1}{2}$ hour studying English, and 3 hours studying citizenship.

_____

_____

P   **7.** Thirty percent of the committee members voted in favor of budget cuts. Twenty percent voted against the cuts. There are 120 members on the committee.

_____

_____

M   **8.** Jen drove 130 miles in 2 hours. Sean drove 150 miles in 3 hours.

_____

_____

GE   **9.** A new parking lot is to be paved behind a shopping mall. The lot is 100 yards long and 36 yards wide.

_____

_____

WN   **10.** Fernando worked 8 hours on Monday, 9 hours on Tuesday, 8 hours on Thursday, and 10 hours on Saturday.

_____

_____

Step 2: Find the Information

Zena is looking at apartments to rent. One is $890 per month, including utilities. Another is $775 per month, but the tenant must pay for heat. If heating costs run about $60 per month in the second apartment, how much more can Zena expect to pay for the first apartment?

Underline any information that you need to solve this problem. Do not solve it.

Word problems do more than just ask a question. They also give you information—information that you need to solve the problem.

To solve this problem you need to know

- the rent for the first apartment: <u>$890</u>

- the rent **and** the heating costs for the second apartment: <u>$775</u> and <u>$60</u>.

Remember that the information in the problem includes **labels** as well as numbers. The numbers 890, 775, and 60 refer to *dollars,* not to inches, dogs, or miles per hour.

The chapter called Finding the Information will give you strategies and practice in finding and using information, including problems that have too much or too little information.

Read each problem below and write the information that is needed to solve the problem. Don't forget to include important labels. Do not solve the problems.

EXAMPLE   Jane drove 17 miles at 66 miles per hour. When she saw a police car, she slowed down to 55 miles per hour for the remaining 4 miles to her exit. How long did it take Jane to drive those 21 miles?

Information needed: *17 miles at 66 mph*

and 4 miles at 55 mph

WN 1. Gary's job is to load canned goods into cardboard boxes. The last box he packed held cans containing 32 ounces of tomatoes, 8 ounces of pears, 16 ounces of applesauce, 16 ounces of grapefruit juice, and 32 ounces of kidney beans. What was the total number of ounces contained in this box?

Information needed: _____

_____

D 2. Maria took her daughter out for lunch. They bought 2 cheeseburgers, 1 small fries, 1 large fries, and 2 medium colas. Before tax, how much did the food cost?

Information needed: _____

_____

DRIVE-IN BURGERS	
HAMBURGER	$1.25
CHEESEBURGER	$1.75
FRENCH FRIES	$.75/.95
SOFT DRINKS	$.50/.75/1.00
CONES	$.75

D 3. Robert was able to drive 211.4 miles on his last tank of gas. How many miles per gallon did he get if the tank held 12.1 gallons of gas?

Information needed: _____

_____

F 4. A candy recipe calls for $1\frac{1}{2}$ cups of sugar for every 6 ounces of chocolate. If a chef used 36 ounces of chocolate in the candy, how much sugar did she use?

Information needed: _____

_____

F 5. A carpenter is building a planter from railroad ties. He wants the planter to be $2\frac{1}{2}$ feet high. If he stacks up the railroad ties, and each is 4 inches high, how many ties will he need to build one side?

Information needed: _____

_____

Step 3: Make a Plan

Carmelita has a sheet of dough 18 inches wide to cut into strips. Each strip must be 2 inches wide. How many strips can she get from one sheet of dough?

What calculation do you need to do to solve this problem?

You're right if you decided to *divide* 18 by 2. Did it take you a few minutes to decide whether to multiply or divide? If so, that's okay. You have just discovered a very important part of the problem-solving process—deciding what operation to use.

Remember, there are just four **operations**—addition, subtraction, multiplication, and division. However, many mistakes with word problems are made by choosing the wrong operation—for example, *multiplying* 18 by 2 in the problem above.

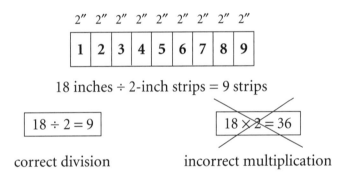

18 inches ÷ 2-inch strips = 9 strips

18 ÷ 2 = 9		18 × 2 = 36
correct division		incorrect multiplication

These mistakes arise from working too quickly or not really understanding what happens to numbers when you perform different operations.

The chapter called Making a Plan will give you strategies and practice in deciding what operations to use to solve a problem.

Read each problem below and decide what operation you need to perform to get the correct answer. Do not solve the problems.

EXAMPLE   A desk that normally sells for $145 has been marked down by $30. What is the sale price of the desk?

Operation: *Subtract: $145 – 30*

WN 1. Roxanne weighs 120 pounds and her daughter weighs 30 pounds. How many times heavier is Roxanne than her daughter?

Operation: _____

WN 2. In Canfield last year, a total of 148 people died from handgun wounds. An additional 320 were injured by handguns. How many people in Canfield either died or were injured by handguns last year?

Operation: _____

D 3. When a driver left his warehouse, he looked at his truck odometer. After he reached his destination, he read it again. How long was the trip?

Before
↑
(at warehouse)

After
↑
(at destination)

Operation: _____

D 4. On a recent drug bust, law enforcement officials seized 110 kilograms of cocaine. If a kilogram equals 2.2 pounds, how many pounds were seized?

Operation: _____

F 5. Saul caught a 3-pound salmon, a $3\frac{1}{2}$-pound bass, and a 2-pound bluefish on his vacation last week. How many pounds of fish in all did he catch?

Operation: _____

Step 4: Solve the Problem

Matthew has 987,475,981$\frac{1}{8}$ yards of rope. He cuts the rope into pieces that measure 561.89341 feet each. How many pieces of rope does he have?

DON'T SOLVE THIS PROBLEM!
What seems especially hard about this problem?

The size and complexity of the numbers make this problem difficult to solve. You'll be happy to know that for this book, and for most books and tests, you will *never* have to work with numbers as large and complex as those above. In most cases, the numbers you work with in your everyday life do not require lengthy calculations either.

Did You Know . . . ?

- The word problems on most mathematics tests do not require complicated and time-consuming computation.

- Most tests are actually trying to determine your problem-solving ability as well as your *basic* computation skills.

- Calculators and computers are doing more and more of the actual computation in the world today, and they help people solve math problems easily and accurately.

These facts may make you feel more at ease with math tests. However, remember that a calculator cannot choose information for you or decide what operation should be used. These "thinking skills," which are so important in word problems, are also the skills important in everyday life.

Knowing how to add, subtract, multiply, and divide accurately is *extremely* important in your work with word problems. So is working with percent, fractions, and decimals. However, this book will not concentrate on these kinds of computations, as there is not space to cover everything. To improve your math, study computation skills while you work through this book.

The chapter called Solving the Problem will concentrate on ways to arrive at a correct solution, and ways to avoid the traps of incorrect solutions.

Practice your computation skills by doing the following problems. Remember to work neatly and carefully.

1. $3,987 + 452 =$

2. $452 \div 4 =$

3. $981 \times 34 =$

4. $9,082 - 28 =$

5. $4\frac{1}{2} + 21\frac{1}{4} =$

6. $4.56 - 3.8 =$

7. 45% of $90 =$

8. $33\frac{5}{6} \times 6 =$

9. 32 is what percent of 50?

10. 66 is 30% of what number?

Step 5: Check Your Answer

Mr. Anton paid an 8% sales tax on his new set of utility shelves. If the price of the shelves was $60 before tax, what total amount did Mr. Anton pay for the shelves?

a. $4.80   **d.** $68.00
b. $48.00   **e.** $480.00
c. $64.80

Michael, a math student, read through this problem and decided he needed to find 8% of $60. He did the math and came up with an answer of $4.80. Michael was working quickly, and because he saw $4.80 listed as an answer choice, he chose **a. $4.80** as the correct answer.

Where did Michael go wrong?

If he had taken time to see if his answer made sense, he probably would have discovered his mistake. He gave the amount of tax, but the problem asks for the *total amount paid.*

$$\begin{array}{ccccc} \$60 & + & \$4.80 & = & \$64.80 \\ \downarrow & & \downarrow & & \downarrow \\ \text{shelves} & & \text{tax} & & \text{total} \end{array}$$

To avoid mistakes like Michael's, make sure that

- you have answered the question *(What total amount did he pay?)*

- your answer makes sense *(How could the total cost—including the tax—be less than the cost of the shelves alone?)*

Many students stop at *Step 4: Solve the Problem.* Once they have an answer, they assume it is *the right answer.* This is especially true on multiple-choice tests, where a number "on the way to the answer" may be listed among the answer choices. You can avoid errors like these if you take time to *check your answer.*

The chapter called Solving the Problem will give you strategies and practice in checking your answer to see that it answers the question asked and is sensible.

Each problem below has two possible answer choices. You don't need to compute to solve the problems. You should be able to see which answer choice is correct simply by deciding which choice answers the question asked and is sensible. Circle the correct choice.

WN 1. Anthony lost 14 pounds, and he now weights 167 pounds. What was his weight before he lost the 14 pounds?

    a. 181 pounds   **b.** 153 pounds

WN 2. The Myersons drove at a speed of 65 mph for 130 miles. How many hours were they driving?

    a. 65 miles   **b.** 2 hours

WN 3. To save time, Tim had 3 friends help him deliver fliers. If each of the 4 boys delivered 72 fliers, how many fliers were delivered in all?

    a. 24 fliers   **b.** 288 fliers

M 4. Angel added 4 cups of water to 2 cups of juice concentrate. How many pints of liquid did she have then?

    a. 6 cups   **b.** 3 pints

M 5. A bus driver covered 195 miles on one tank of gas. The gas tank holds 15 gallons. How many miles per gallon did he get on that tank?

    a. 13 miles   **b.** 180 gallons

F 6. Jake lost $47\frac{1}{2}$ pounds. His dad, David, lost twice that amount. How many pounds did David lose?

    a. $23\frac{3}{4}$ pounds   **b.** 95 pounds

P 7. Sarah answered 80% of her math test questions correctly. The test had 15 questions on it. How many did she get right?

    a. 12 questions   **b.** 18 questions

GE 8. How many inches of lace will Eliza need to trim a square pillow with a side that measures 6 inches?

    a. $1\frac{1}{2}$ inches   **b.** 24 inches

UNDERSTANDING THE QUESTION

Read Carefully

Problem 1
The three top prizes in an essay contest were $400, $200, and $50. What was the difference between the two top prizes?

Problem 2
The three top prizes in an essay contest were $400, $200, and $50. What was the total given away in prizes?

Are the two problems above the same? They both refer to an essay contest, and the information provided for you is exactly the same. But the questions are not the same.

The two problems show the importance of Step 1 of the problem-solving process: **understand the question.** Before you can do anything with the numbers given in a problem, you must understand what you are being asked to find.

What is the answer to each problem?
What did you have to do to get each answer?

You're right if you said the answers are **$200** and **$650.**

The first problem:

$400 − 200 = **$200**

The second problem:

$400 + 200 + 50 = **$650**

Even though the information in the problems is the same, the answers are different because the questions are different.

Try the next exercise. It will help you practice understanding what you are being asked to find.

 Each set of problems gives the same information but asks different questions. Pay special attention to each question as you solve the problems.

WN 1. **a.** On a recent trip together, Nancy drove 70 miles, Dave drove 45 miles, and Louis drove 95 miles. How many more miles did Nancy drive than Dave?

WN   **b.** On a recent trip together, Nancy drove 70 miles, Dave drove 45 miles, and Louis drove 95 miles. How many miles did the three drive in all?

WN   **c.** On a recent trip together, Nancy drove 70 miles, Dave drove 45 miles, and Louis drove 95 miles. What was the average number of miles driven by the three?

WN   **d.** On a recent trip together, Nancy drove 70 miles, Dave drove 45 miles, and Louis drove 95 miles. If their total trip was going to be 250 miles, how many more miles did they still have to go?

GE **2.** **a.** A rectangular construction site is shown below. If a fencing company is to enclose the entire site, how many yards of fence will it need?

GE   **b.** A rectangular construction site is shown below. Find the number of square yards in this site.

WN   **c.** A rectangular construction site is shown below. How many times longer is the site than it is wide?

D   **d.** A rectangular construction site is shown below. If each square yard was purchased for $1.25, what was the cost of the whole site?

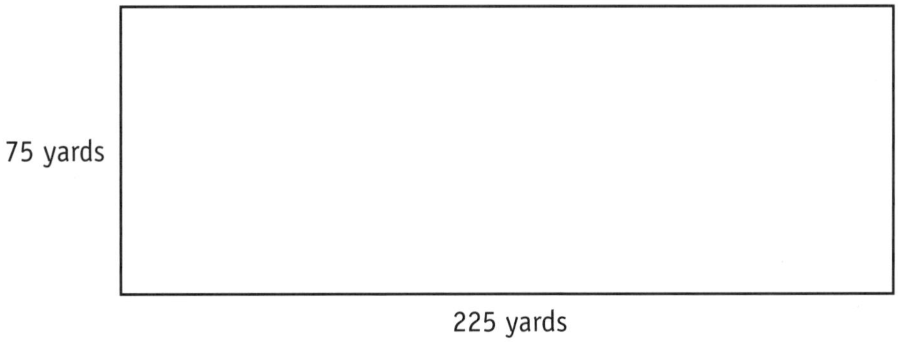

75 yards

225 yards

You are going to help write some word problems. Read the information and then write two questions based on the information. Finally, solve the problems.

EXAMPLE   Nedra worked for 13 hours over the weekend. Last weekend she worked only 5 hours.

Question 1: *How many more hours did she work this weekend than last?*

Solution: 13 − 5 = **8 hours**

Question 2: *How many hours in all did she work over the 2 weekends?*

Solution: 13 + 5 = **18 hours**

3. Both Mr. and Mrs. Phillips are employed. Mr. Phillips takes home $2,700 each month, while his wife takes home $3,000.

Question 1: _____

_____

Solution:

Question 2: _____

_____

Solution:

4. Three friends went out to dinner one evening. Gwen had a chef's salad for $5.50, Michael had pot roast for $7.00, and Robert had a hamburger for $4.25. They all split an ice cream sundae for $2.00.

Question 1: _____

_____

Solution:

Question 2: _____

_____

Solution:

5. Sue and Joe Murphy spent 4 hours at the racetrack. During that time, Sue won $160 and Joe lost $200.

   Question 1: _____

   _____

   Solution:

   Question 2: _____

   _____
   (Hint: Think about each person's winnings per hour.)
   Solution:

6. The new soccer field measures 45 yards long and 25 yards wide.

   Question 1: _____

   Solution:

   Question 2: _____

   Solution:

7. Sandy spent 45 minutes driving to the shopping mall. She went to 3 stores and spent 10 minutes in each.

   Question 1: _____

   Solution:

   Question 2: _____

   Solution:

8. Vikki wrote a 10-page paper in 2 days. Hannah wrote a 12-page paper in 3 days. Maggie wrote an 8-page paper in 4 days.

   Question 1: _____

   Solution:

   Question 2: _____

   Solution:

9. Mrs. Kirby spent $200 on a new suit. Ms. Welch found the same suit for $50 less.

   Question 1: _____

   Solution:

   Question 2: _____

   Solution:

10. A carpenter cut the three boards shown below.

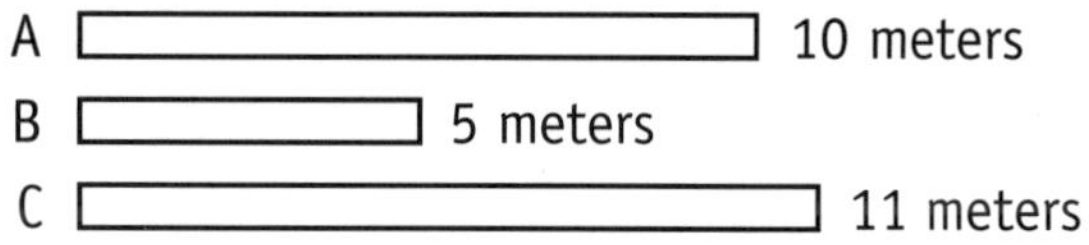

    A ⬜⬜⬜⬜⬜⬜⬜ 10 meters
    B ⬜⬜⬜⬜ 5 meters
    C ⬜⬜⬜⬜⬜⬜⬜⬜ 11 meters

    Question 1: _____

    Solution:

    Question 2: _____

    Solution:

11. Joanna worked 14 hours on Monday, 8 hours on Tuesday, and 10 hours on Wednesday.

    Question 1: _____

    Solution:

    Question 2: _____

    Solution:

12. Doug and Sam spent 3 hours in a video arcade. Doug used up 96 tokens, and Sam used 81.

    Question 1: _____

    Solution:

    Question 2: _____

    Solution:

Some Tricky Questions

Denise bought some lace to sew around 3 square pillows. Each pillow measured 4 inches on a side. How many *feet* of lace did Denise need to trim all 3 pillows?

a. 3   **b.** 4   **c.** 12   **d.** 16   **e.** 48

Which is the correct answer to the problem above?

If you answered choice **b. 4,** you answered correctly. However, if you answered choice **e. 48,** you have stumbled on a common error made in word problems. **You did not answer the question that was asked.**

What steps are necessary to solve the problem?

STEP 1

$$\begin{array}{r} 4 \text{ inches} \\ \times\ 4 \text{ sides} \\ \hline 16 \text{ inches} \end{array}$$

STEP 2

$$\begin{array}{r} 16 \text{ inches} \\ \times\ 3 \text{ pillows} \\ \hline 48 \text{ inches} \end{array}$$

> **TIP**
> On some tests a word problem involving measurements may have a measurement word underlined or in *italics*. (The problem above had *feet* in italics.) Be alert for this clue that you may need to change measurement units.

But reread the question. The question asks for the total number of *feet* of lace. Therefore, you need another step.

STEP 3

$$12\overline{)48 \text{ inches}} = 4 \text{ feet}$$

inches in 1 foot

Notice how important it is to read the question carefully!

To Be Test Wise, Know These Equivalencies

12 in. = 1 ft	16 oz = 1 lb	60 sec = 1 min	2 c = 1 pt
3 ft = 1 yd	2,000 lb = 1 ton	60 min = 1 hr	2 pt = 1 qt
5,280 ft = 1 mi	1,760 yd = 1 mi	24 hr = 1 day	4 qt = 1 gallon

 Solve the following word problems. Be sure you answer the question asked.

WN 1. Celeste wants to know how long it will take her to walk from her apartment to a department store that is 4 miles away. If she walks 1 mile in 15 minutes, how many *hours* will it take her to get to the store?

    a. 1   **b.** 2   **c.** 11   **d.** 30   **e.** 60

WN 2. Celeste wants to know how long it will take her to walk from her apartment to a department store that is 4 miles away. If she walks 1 mile in 15 minutes, how many *minutes* will it take Celeste to get to the store?

    a. 1   **b.** 2   **c.** 11   **d.** 30   **e.** 60

WN 3. Each small truck from the Ace Removal Company can haul 500 pounds of trash at a time. On Tuesday the company has jobs to remove approximately 1,500 pounds of trash from one construction site, 500 pounds from another site, and 2,500 pounds from a third site. How many truckloads in all will Ace remove?

    a. 3   **b.** 6   **c.** 9   **d.** 900   **e.** 4,500

WN 4. Each small truck from the Ace Removal Company can haul 500 pounds of trash at a time. On Tuesday the company has jobs to remove approximately 1,500 pounds of trash from one construction site, 500 pounds from another site, and 2,500 pounds from a third site. How may pounds in all will Ace remove?

    a. 3   **b.** 6   **c.** 9   **d.** 900   **e.** 4,500

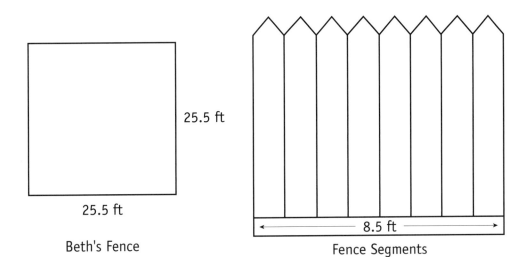

25.5 ft

25.5 ft

Beth's Fence

8.5 ft

Fence Segments

D 5. The fence that Beth needs to build is pictured above. The fencing is sold in segments as shown and is priced at $4.70 per foot. How many segments of fence will Beth have to buy?

a. 8.5 **b.** 12 **c.** 56.4 **d.** 102 **e.** 479.4

D 6. The fence that Beth needs to build is pictured above. The fencing is sold in segments as shown and is priced at $4.70 per foot. How much money will Beth have to spend on the fence?

a. $8.50 **b.** $12.00 **c.** $56.40 **d.** $102.00 **e.** $479.40

F 7. A standard running track is 440 yards around, or $\frac{1}{4}$ mile. At a recent track meet, Danita ran a 440-yard race and a 1-mile race. How many *yards* did she run in all?

a. $1\frac{1}{4}$ **b.** 441 **c.** 1,440 **d.** 1,760 **e.** 2,200

F 8. A standard running track is 440 yards around, or $\frac{1}{4}$ mile. At a recent track meet, Danita ran a 440-yard race and a 1-mile race. How many *miles* did she run in all?

a. $1\frac{1}{4}$ **b.** 441 **c.** 1,440 **d.** 1,760 **e.** 2,200

Working with Set-Up Questions

Maude earns $56 per day during the week and $80 per day on weekends. Which expression gives you the amount of money Maude earned on Monday and Saturday?

a. $56 + $80

b. $80 − $56

c. 2 × $56

d. $80 × $56

e. $80 ÷ $56

How does this problem look different from the other word problems you have seen?
What is the question asking you to find?

Read the problem again **without reading the answer choices.** Does the problem seem easier now? You'll probably be relived to learn that once you know *how* to solve the problem, you'll be able to choose the correct answer on a question like this. The problem above is a type found on many math tests. It is often called a **set-up question** because you have to choose one way that a problem could be set up for solution.

As you know, the first step in solving *any* word problem is understanding what you are being asked to find.

What are you being asked to find in this problem?

In this problem, you are being asked to find an **arithmetic expression.** In mathematics, an expression is a way of showing how a solution can be found. An expression is made up of two or more numbers and one or more signs of operation (+, −, ×, ÷).

Look at the problem on the top of page 44. How can you find out the amount of money Maude earned on those two days?

$$\$56 \quad + \quad \$80 \quad = \quad \text{total}$$
$$\uparrow \qquad\qquad \uparrow$$
$$\text{Monday} \qquad \text{Saturday}$$

If you look at the answer choices listed, **a. $56 + $80** gives you the correct expression. The other choices will not tell you how much Maude earned.

Before you move on to more challenging set-up problems, try the next exercise, which will give you more practice in working with arithmetic expressions.

....

Choose the correct arithmetic expression for each of the following word problems. Do not solve the problems.

WN 1. Carl needs an additional $240 to cover the security deposit on the apartment he wants. He has already put aside $700 for the deposit. Which expression shows the total amount of the deposit?

    a. $700 \div 240$ **d.** 240×700

    b. $700 - 240$ **e.** $240 \div 700$

    c. $240 + 700$

WN 2. The office suite where Sondra works uses 3,000 kilowatt-hours of electricity each month. Which expression shows the number of kilowatt-hours used in a 3-month period?

    a. $3{,}000 + 3$ **d.** $3{,}000 - 3$

    b. $3{,}000 \times 3$ **e.** $3{,}000 \div 12$

    c. $3{,}000 \div 3$

WN 3. The temperature in Springvale yesterday was 64 degrees. Today it is 59 degrees. Which expression shows the number of degrees cooler it is today?

    a. $64 + 59$ **d.** $59 + 5$

    b. $64 - 5$ **e.** $64 - 59$

    c. $64 \div 59$

WN **4.** The computer at the research center can print about 8 pages per minute. Mr. Lopez has a 70-page document to print. Which expression shows approximately the number of minutes Mr. Lopez should allow for the printing?

a. 70×8 **d.** $70 \div 60$
b. $70 \div 8$ **e.** $60 \div 8$
c. $70 - 8$

P **5.** A clerk at Mama's Deli wants to figure out the tax on an order of $7.10. If the food tax is 6 percent, which expression represents the tax on that order?

a. $\$7.10 + 0.06$ **d.** $\$7.10 - 0.06$
b. $\$7.10 - 0.06$ **e.** $0.06 - \$7.10$
c. $\$7.10 \times 0.06$

GE **6.** Which expression shows the area of Jim's garden in square feet?

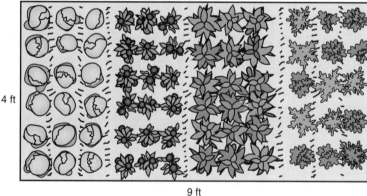

4 ft

9 ft

a. $4 + 9$ **d.** $9 \div 4$
b. $4 + 4$ **e.** 9×4
c. 4×4

More on Set-Up Problems

Fifteen percent of all participants in a consumer survey said they preferred rice to potatoes. Eleven men and 25 women participated. Which expression shows the number of people in the survey who preferred rice?

a. $15 + 11 + 25$
b. $15(11 + 25)$
c. $(0.15 \times 11) + 25$
d. $(0.15 \times 25) + 11$
e. $0.15(11 + 25)$

This set-up problem may seem more complicated than the ones you worked with in the last exercise. As you can see, you need more than one operation to get the answer. Therefore, the answer choices listed are a little more complicated. Let's take a look at some steps in choosing the correct expression.

STEP 1	Talk about or "picture" the process needed to solve a problem. Make a statement that tells what needs to be done.	*I need to add men to women and find a percent of that total.*
STEP 2	Now fit the numbers given in the problem into your statement.	*I need to add 11 to 25 and then take 15% of that total.*
STEP 3	Write an expression using numbers and operation symbols (before you look at the answer choice).	*0.15(11 + 25) 0.15 means 15%*
STEP 4	Find the answer choice that matches the one you have written.	*e. 0.15(11 + 25)*

Using Parentheses ( )

Were any of the operation symbols new to you? Notice that the answer choices included parentheses ( ). Parentheses mean two things in arithmetic expressions.

1. **Do this first.** Any computation inside parentheses should be done before any operation outside.

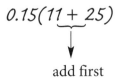

0.15(11 + 25)

add first

2. **Multiply.** When a number is written outside a pair of parentheses, you should multiply.

$$0.15(11 + 25)$$
means
$$0.15 \times (36)$$

Using the Division Bar (—) or Slash Mark (/)

Look at another example of a set-up problem.

On Monday 291 books were checked out of the town library. On Tuesday 145 were checked out, and on Wednesday 202 books were checked out. Which expression tells the average number of books taken out over the 3 days?

a. $291 + 145 + 202$

b. $\dfrac{202 + 145}{2}$

c. $\dfrac{291 + 145 + 202}{3}$

d. $\dfrac{291 + 145 + 202}{2}$

e. $201 - 202 - 145$

> **TIP**
> In some cases, an expression for a problem can be written in a number of ways and still have the same value.
>
> For example, these all have the same value:
> $0.15(11) + 0.15(25)$
> $0.15(25) + 0.15(11)$
> $0.15(25 + 11)$
>
> If you are confused by a set-up problem, think about the different ways that some expressions can be written.

STEP 1	Figure out or "picture" the problem.	*I need to add the numbers to be averaged, then divide by how many numbers I am averaging.*
STEP 2	Fill in the numbers.	*I need to add 291, 145, and 202, then divide by 3.*
STEP 3	Write an expression.	$\dfrac{291 + 145 + 202}{3}$
STEP 4	Find the matching answer.	*c.* $\dfrac{291 + 145 + 202}{3}$

Did you notice how division was indicated in the problem above? A **division bar** (or fraction bar) was used.

$$\frac{291 + 405 + 202}{3} \longleftarrow \text{this bar means divide}$$

In some cases you may see a **slash** (/) instead of the division bar. The problem above could also be written as

$(291 + 405 + 202) / 3$

**Write the correct expression for each of the following problems.
Do not solve the problems.**

WN 1. Grady drove at a speed of 60 mph for the first 3 hours of a trip.
Because he was late, he drove the last 2 hours at a rate of 65 mph.
Write an expression that shows the number of miles Grady drove
on this trip.

Expression: $(60 \times 3) + ($ ____ $\times$ ____ $)$

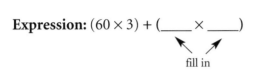

fill in

WN 2. A science class made up of 12 girls and 15 boys went on a field trip
one day. The children were divided equally into three groups, each
group with its own adult supervisor. Write an expression that
shows how many students were in each group.

Expression: _____

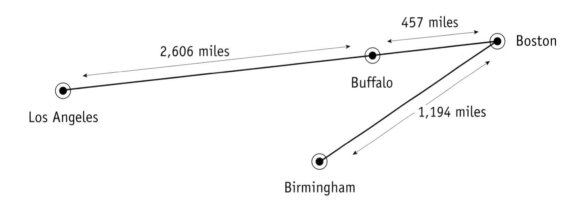

457 miles

2,606 miles

Boston

Buffalo

1,194 miles

Los Angeles

Birmingham

WN 3. A salesman traveled from Birmingham to Boston, then from
Boston to Buffalo in one day. The following day he flew from
Buffalo to Los Angeles. Based on the map above, write an
expression to show the difference in miles traveled those two days.

Expression: _____

D 4. For a batch of her famous spaghetti sauce, Mrs. Mason bought
4 cans of plum tomatoes at $1.29 per can and 4 cans of stewed
tomatoes at $0.89 per can. Write an expression that shows how
much Mrs. Mason paid for these ingredients.

Expression: _____

P 5. The Terrific Tee Shop has a sale of 35% off all T-shirts in the store. The Mugford family came in and bought two shirts originally priced at $12.99. Write an expression that shows how much money the Mugfords *saved* during this sale.

Expression: _____

GE 6. Juan and Miriam wanted to put a rope fence around the new grass they had just planted. The sides of the triangular area measured 4 feet, 9 feet, and 16 feet. Write an expression that shows how much they spent on rope if each foot of rope costs $0.30.

Expression: _____

Choose the correct arithmetic expression for each problem. Do not solve the problems.

WN 7. The members of David's bicycle club rode 13 miles on Thursday, 11 miles on Friday, 15 miles on Saturday, and 10 miles on Sunday. Which expression shows the *average* number of miles each member of the club rode on these days?

    a. $13 + 11 + 15 + 10$ **d.** $4(13 + 11 + 15 + 10)$

    b. $\dfrac{13+11+15+10}{2}$ **e.** $2(13 + 11 + 15 + 10)$

    c. $\dfrac{13+11+15+10}{4}$

WN 8. Most doctors recommend that adults consume no more than 3,300 milligrams of sodium per day. This morning Mr. Shreve ate two muffins, each containing 390 milligrams of sodium. If Mr. Shreve wants to stay within the recommended guidelines, which expression shows the maximum milligrams of sodium he may consume for the rest of the day?

    a. $3{,}300 - 390$ **d.** $3{,}300 + (2 \times 390)$

    b. $3{,}300 + 390$ **e.** $3{,}300 - (2 \times 390)$

    c. $\dfrac{3{,}300}{2 \times 390}$

WN 9. During the months before it went out of business, Powell's Party Supplies experienced a severe decrease in sales. Based on the graph at the right, which expression shows the difference in dollar sales between July and October?

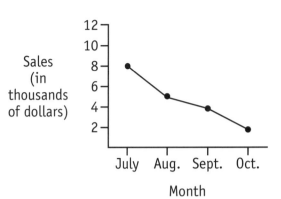

Powell's Party Supplies

Sales (in thousands of dollars)

Month

a. $8 - 5 - 4 - 2$
b. $1,000(8 - 5 - 4 - 2)$
c. $8 - 2$
d. $1,000(8 + 2)$
e. $1,000(8 - 2)$

WN 10. Maureen withdrew $50 from her savings account on Thursday afternoon. On Friday morning she deposited $110. If Maureen's account balance on Thursday morning had been $890, which expression shows her account balance on Friday afternoon?

a. $890 - (\$50 + \$110)$
b. $(\$890 - \$50) + \$110$
c. $890 - \$50 - \$110$
d. $50 + \$110 + \$890$
e. $50 - \$110 - \$890$

F 11. The population of India in 1999 was approximately 990,000,000 people. The population of the United States was about $\frac{1}{3}$ that number. Which expression shows the combined population of India and the United States?

a. $\frac{1}{3} + 990,000,000$
b. $\frac{1}{3}(990,000,000)$
c. $\frac{1}{3}(990,000,000) + 990,000,000$
d. $3(990,000,000) + 990,000,000$
e. $3(990,000,000)$

GE 12. A carpet installer wants to cover $\frac{1}{3}$ of a rec room floor with industrial-strength carpet. The rest of the floor will be tile. If the floor is 10 feet wide and 24 feet long, which expression represents the number of square feet of carpet needed?

a. $10 \times 24 \times 3$
b. $\frac{1}{3}(10 \times 24)$
c. $\frac{1}{3}(10 + 24)$
d. $3(10 \times 24)$
e. $\dfrac{3}{10 \times 24}$

Mixed Review

Choose the correct answers for the following problems.

WN 1. A train traveled at 100 mph for 2 hours, then went an additional 2 hours at 85 mph. Which expression gives the number of miles the train traveled?

   a. $100 + 85$ **d.** $(2 \times 100) + (2 \times 85)$

   b. $4(100 + 85)$ **e.** $\dfrac{2 \times 100}{2 \times 85}$

   c. $\dfrac{100}{2} + \dfrac{85}{2}$

WN 2. Last year 2,050 law enforcement officers were employed in Miami, Florida. In addition, there were 405 civilian law enforcement employees. The total number of law enforcement personnel was down 27 people from the previous year. Which expression shows the number of law enforcement employees last year?

   a. $2{,}050 - 27$ **d.** $2{,}050 + 27$

   b. $(2{,}050 + 405) + 27$ **e.** $27(2{,}050 + 405)$

   c. $(2{,}050 + 405) - 27$

WN 3. Sara's time sheet is shown below. How many shifts did she work if a shift is 4 hours?

   a. 4
   b. 6
   c. 12
   d. 48
   e. 52

Sarah Hazelton
Employee 2213

Day	Hours
Monday	4
Tuesday	8
Wednesday	8
Thursday	12
Friday	12
Saturday	4

D 4. Rafael jogs 6 nights per week. The calendar where he records his mileage is shown at the right. Find the average number of miles he ran Monday through Saturday.

a. 5.4
b. 6.0
c. 9.0
d. 18.4
e. 36.0

S	M	T	W	T	F	S
1 O	2 6.2	3 4.2	4 6.2	5 5.4	6 6.2	7 7.8
8	9	10	11	12	13	14
15	16	17	18	19	20	21
22	23	24	25	26	27	28
29	30	31				

P 5. A total of 12,000 people are expected to see the new museum exhibit. Eight thousand of these people will be coming on tour buses. What percent of the people will be from the tour buses?

a. 8%
b. 12%
c. $33\frac{1}{3}\%$

d. $66\frac{2}{3}\%$
e. 100%

P 6. A customer at Davio's bought two suits priced at $375 each. If the clothing tax was 7%, which expression shows the total amount the customer paid for the suits?

a. $0.07 \times \$375$
b. $(2 \times \$375) + 0.7(2 \times \$375)$
c. $(2 \times \$375) + 0.07(2 \times \$375)$

d. $0.17(2 \times \$375)$
e. $0.07(2 \times \$375)$

P 7. Mr. and Mrs. Stern sold their home for $140,000. Out of that money, they had to pay the realtor a 6% commission. They took the balance of the money and put $30,000 down on a new house. How much money could they then put in the bank?

a. $8,400
b. $84,000
c. $101,600

d. $110,000
e. $131,600

M 8. If a tailor cuts a 25-foot piece of fabric into strips that measure 4 *inches* wide, how many strips of fabric will he have?

   a. $6\frac{1}{4}$ **d.** 100

   b. 8   **e.** 1,200

   c. 75

M 9. Beverly poured 4 cups of flour, 2 cups of sugar, and 6 cups of milk into a bowl. How many total *pints* did she mix together? (Use the equivalencies chart on page 42 if necessary.)

   a. 4   **d.** 12

   b. 5   **e.** 18

   c. 6

GE 10. Sketches of Ellen's gardens are shown below. The insect killer she plans to use covers 10 square *yards* per jar. How many jars does Ellen need to buy?

   a. 4.5   **d.** 9.5

   b. 5.0   **e.** 45.0

   c. 9.0

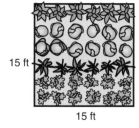

FINDING THE INFORMATION

Looking at Labels

Mrs. Martínez purchased 4 oranges, 3 pears, 6 pounds of ground beef, and 5 red apples. How many pieces of fruit did she buy in all?

If you read this problem quickly and just add up all the numbers, you won't get the right answer. To solve this and other word problems correctly, you must pay attention to **labels.**

What are the labels in the problem above?
Which of the following is the correct solution?

Solution 1	Solution 2
4 oranges	4 oranges
3 pears	3 pears
6 pounds of beef	+ 5 apples
+ 5 apples	12
18	

The words *oranges, pears, ground beef,* and *apples* are important **labels** in this problem.

What label should your answer have?

The problem asks for the number of pieces of *fruit;* therefore, you should add up the number of apples, pears, and oranges. Ground beef is not a fruit, so it should not be added in the total.

12 pieces of fruit is the correct answer.

For every number in the word problems below, circle the label that belongs with it. Then solve each problem, *including a label in your answer.*

EXAMPLE   Ron divided his classroom into 5 (groups.) Each group contained 6 (students.) How many students were in Ron's classroom?

$$5 \text{ groups} \times 6 \text{ students} = 30 \text{ (students)}$$

WN **1.** The orchard Nancy wanted to purchase contained 11 acres of apple trees. Approximately 120 trees grew on each acre. How many trees did the orchard contain?

WN **2.** On Dan Brunner's used-car lot there are 4 Chevrolets, 10 Fords, 10 Chryslers, and 9 Toyotas. How many cars are on Dan's lot altogether?

F **3.** On Vince's recent airplane trip he covered 900 miles in $2\frac{1}{2}$ hours. On average, how many miles per hour did the airplane fly?

M **4.** If 1 meter equals approximately 39 inches, about how many inches are in 10 meters?

GE **5.** For the room shown below, how many square feet of linoleum would you need to cover the floor?

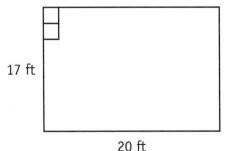

17 ft

20 ft

Finding Hidden Information

During a particularly rainy April, Charleston received 3.5 inches of rain in 1 week. What was the average number of inches per day?

Sometimes information in a word problem can be hidden; in other words, the numbers needed to solve the problem might not be obvious.

Take a minute to think out loud about how you might solve the problem above.

STEP 1 Perhaps you said something like this: *I will divide the total number of inches by the number of days in a week to get the average inches per day.*

STEP 2 Then you put in the numbers. *I need to divide 3.5 by 7.*

STEP 3 Next, do the computation. *3.5 ÷ 7 = 0.5 inches*

Where did the number 7 come from? It wasn't in the problem, was it?

Charleston received 3.5 inches of rain in 1 week.   1 week = 7 days

You needed to remember that there are 7 days in a week and to use that information in the problem.

Take a look at another word problem with hidden information.

A mail carrier on the morning route delivered letters to ninety houses. Her partner on the afternoon route delivered to twice that many. How many houses in all did the two mail carriers deliver to?

Is this a math problem?
Where are the numbers?

Don't give up—this problem has plenty of numbers to work with.

- Circle the number of houses the morning carrier delivered to.

- Now circle the clue that tells you how many houses the afternoon carrier delivered to.

Did you circle *ninety* and *twice that many?* This is the information that you need to solve the problem correctly.

$$90 + (2 \times 90) = \textbf{270 houses}$$

morning ⟶ ⎿ afternoon

Take a look at one more word problem with hidden information. Notice how important it is to **read carefully.**

In all but 3 months of 1999, Sasha sold one dictionary per month in her job as a door-to-door salesperson. If Sasha earned $18.50 for each dictionary sold, what were her profits on 1999 dictionary sales?

In how many months did Sasha sell a dictionary? Multiply this number by $18.50.

If you thought Sasha sold 3 dictionaries, you did not find the hidden information in this problem. If she sold dictionaries *all but 3 months,* for how many months did she sell dictionaries?

```
  12 months
-  3 months without sale
   9 months with sales
```

Sasha sold one dictionary each of the 9 months.
Since $9 \times \$18.50 = \$166.50$, Sasha earned **$166.50.**

What was the hidden information in this problem?

First of all, you needed to know what the phrase *all but* meant; second, you needed to know that there are 12 months in a year.

> **TIP**
> Read all the information carefully to see if a problem has a number in word form (*eight*) or a hidden number (*1 week = 7 days*).

Solve the following word problems. Read carefully and watch out for hidden information.

WN 1. Richard's monthly salary is twice what he pays for room and board at the Daze Inn. If he pays $680 per month at the inn, what is his monthly salary?

WN 2. Wanda walked 15 blocks from her home to the coffee shop, then walked the same distance back. If she walks at a rate of 15 blocks per hour, how much time did she spend walking?

WN 3. Mary did a survey of her coworkers to see how they used the lounge at work. She found that 3 people took naps, 8 people came in to talk with coworkers, 17 people came in to smoke cigarettes, and 10 people bought coffee. How many people in all used the lounge that day?

WN 4. A cook at Franco's Café baked all but 10 pounds of the potatoes he had on hand for Saturday night dinners. He started out with 5 bags of potatoes, each weighing 8 pounds. How many pounds of potatoes did the cook bake for Saturday night?

D 5. For an art project, Francine buys 3 sheets of red felt, 10 sheets of yellow felt, and a dozen sheets of black. If each sheet of felt costs $0.55, how much does Francine spend on felt?

D 6. At the Redville Cinema, one hundred eighteen people each paid $7.00 to see the early evening show. Seventy-two people paid the same amount to see the late show. How much money did the cinema take in for these two shows?

P 7. Sixty-four percent of all the dairy cows on U.S. farms are located in the Midwest. Based on the graph at the right, how many dairy cows were in the Midwest in the year shown?

Livestock on U.S. Farms

M 8. Belinda works hard to keep her commitment as a caseworker and as a good parent. She leaves for work at 8:00 in the morning and gets home at 6:00 in the evening. She finds that 6 hours of sleep per night is plenty for her. How many hours does this leave for herself and her family?

M 9. To fence his garden, Julius put a stake in the ground every three feet. Then he tied string between the stakes. If Julius's garden is ten *yards* around, how many poles did he need?

GE 10. Max is building a patio for Mrs. López. He will charge her $6 per square foot of patio, including all labor and materials. If Mrs. López wants a patio as shown, how much will Max charge her?

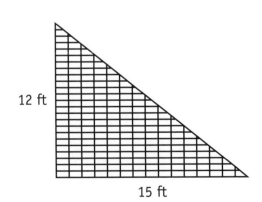

12 ft

15 ft

Extra Information

A salesman was paid $9.50 per hour, plus a bonus of $6.00 for every sale he made. If the salesman worked 10 hours and made 12 sales, how much did he get as a bonus?

a. $60　　**b.** $72　　**c.** $114　　**d.** $167　　**e.** $190

For this problem, review the first two steps of the five-step process on page 24.

STEP 1　What is the question asking?　*It is asking how big a bonus the salesman got.*

STEP 2　What information do you need to answer the question?　*You need to know that he got $6.00 per sale and that he made 12 sales.*

What's tricky about this problem? You are given **more information** than you need to solve it.

Question: *How much did he get as a bonus?*

Information Needed: *$6.00 per sale; 12 sales*

Extra Information: *$9.50 per hour salary;*

10 hours worked

What is the answer to the problem?

You're right if you said **b. $72.** If you thought the answer was **d. $167,** go back and read the question again. Remember, you want to find only the bonus—not the total pay.

> **TIP**
> Word problems may contain extra information. The multiple-choice answers will show choices that you could make if you mistakenly used the extra information. Be careful!

Now look at another word problem and decide what information is needed to solve it.

To get to work each day, Eva must walk 10 minutes from her house to the train, take a 25-minute train ride, walk another 5 minutes to the bus, and ride 15 minutes on the bus. How many minutes does Eva spend riding to work each day?

a. 15   **b.** 30   **c.** 35   **d.** 40   **e.** 55

STEP 1	What is the question asking?	*It is asking for the number of minutes spent riding.*
STEP 2	What information do you need?	*You need to know that she spends 25 minutes on the train and 15 on the bus.*
STEP 3	What operation do you use?	*Add 25 + 15.*
STEP 4	What is the answer?	*c. 40*

If you chose **e. 55** as the correct answer, go back and reread the question. You probably used the extra information about Eva's *walking* time. The question asks for *riding* time.

....

For each problem, write the question and the necessary information in your own words. Cross out the extra information. Do not solve the problems.

EXAMPLE   In 1998, the estimated population of Indiana was 5,800,000 people. ~~About 800,000 of those people lived in the state capital, Indianapolis.~~ ◄——— extra information
If the estimated state population was 5,600,000 in 1990, how many more people lived in Indiana in 1998 than in 1990?

Question: *How many more people lived in Indiana in 1998 than in 1990?*

Necessary Information: *5,800,000 and 5,600,000*

WN **1.** Ron is trying to decide whether or not to buy a new car from Milton's Chevrolet dealership. The car he wants has a base price of $14,395 with an extra $1,200 for air-conditioning, $800 for a sunroof, and $400 for rustproofing. If Ron decides to buy the car with a sunroof and rustproofing, what will it cost him?

Question: _____

Necessary Information: _____

WN **2.** Officer Perry stopped a car that was traveling at a speed of 82 miles per hour. Three hours later he spotted the same car traveling at 70 mph. If the speed limit was 55 mph, by how much had the driver exceeded the limit the first time he was stopped?

Question: _____

Necessary Information: _____

D **3.** An adult's ticket to the carnival is $5.50. A child's ticket is $4.00. If 100 children and twice as many adults paid for the carnival on Sunday, how much money was collected for adult tickets?

Question: _____

Necessary Information: _____

P **4.** Of the 22,500 ice cream cones served by Littlefield's Dairy Bar last summer, 40% were small and 35% were large. If each large cone cost $1.90, how much money did the dairy bar take in for large cones?

Question: _____

Necessary Information: _____

Cross out the extra information and solve the problems.

WN **5.** The Arzadon family drove 396 miles in 7 hours. They used 12 gallons of gas. How many miles per gallon did their car get on this trip?

WN 6. The waitresses at Flannigan's put all their tips into one community pool, then divide the tips equally among all waitresses. One day Sandra earned $45.15 in wages and $51.00 in tips. Carol earned $32.00 in tips that day, while Madge earned $64.00 in tips and $60.05 in wages. How much did each waitress take home in tips?

WN 7. How many more grams of fat does a tablespoon of mayonnaise have than a tablespoon of sour cream?

	Total Fat (grams per tbsp)	Cholesterol (milligrams per tbsp)
butter	11	31
margarine	11	0
mayonnaise	11	8
sour cream	3	5

F 8. Carmella enjoys watching three movies on television and two at the theater each week. Her son watches twice as many movies on television but only half as many in the theater. How many movies does Carmella's son watch in the theater each week?

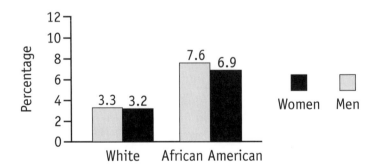

1997 U.S. Unemployment Adults 25 and Over

GR 9. Based on the graph above, how much lower was the unemployment rate for African American women than African American men?

Not Enough Information

At Anchor Hardware, 2-inch nails are sold by the box. A carpenter came in to purchase 4 pounds of these nails. How much did he spend if a box of nails is priced at $2?

a. $2   **d.** $16

b. $4   **e.** not enough information given

c. $8

> **Did you have trouble finding the answer?**
> **What information is missing?**

   Notice that the problem tells you how much a *box* costs—but not how much a *pound* costs. You do not know how many pounds are in a box. Which answer is correct?

   You're right if you said **e. not enough information given.** This type of word problem appears on many tests, and you'll get practice with it throughout this book. Try another one:

Nathan's salary is $20,000 per year. The company he works for withholds 15% of this salary for state tax, federal tax, social security, and health insurance. How much did Nathan pay for health insurance this year?

a. $300   **d.** $9,000

b. $3,000   **e.** not enough information given

c. $6,000

> **What information is missing from the problem?**

   Although you are told the total percentage that is withheld from Nathan's salary, you do not know what part of that goes toward health insurance. If you chose answer **b. $3,000,** you made a common error— you found 15% of $20,000, but you did not answer the question "How much did Nathan pay for *health insurance?*"

Take one more look at a challenging word problem. Don't be fooled by answer choice **e.**!

A total of 25 cheese labels are printed every 3 minutes on Line A at a factory. Line B prints 740 labels every hour. When both lines run nonstop, how many more labels are printed on Line B than Line A in 1 hour?

a. 240   **d.** 715
b. 500   **e.** not enough information given
c. 665

At first you might be tempted to choose **e.** as the answer. But if you work carefully, you'll see that you *are* given enough information. Use the five-step process.

STEP 1	What is the question asking?	*How many more labels are printed on line B than line A in one hour?*
STEP 2	What information do you need?	*Line B = 740 labels* *Line A = 25 labels × 20 3-minute periods in 1 hour = 500 labels*
STEP 3	What's the plan?	*To find how many more, you should subtract.*
STEP 4	Do the math.	*740 − 500 = **240 labels***
STEP 5	Check your answer.	***240** is a sensible answer because it is the only choice less than Line A.* ***a. 240** is the correct choice.*

At first you may not have realized that you *can* figure out how many labels are printed in 1 hour on Line A. Although this information is not stated directly, you can use the numbers provided to find the right answer.

....

None of the following word problems can be solved because some information is missing. Read each problem and write down what information would be necessary to solve the problem.

EXAMPLE   A bus traveled 55 mph for a period of 4 hours. It then picked up speed to 65 mph and continued the trip. How far did the bus travel in all?

Missing Information: *number of hours at 65 mph*

WN 1. Steven's exercise class is full on Tuesdays and Thursdays. On Wednesdays, he gets half as many participants, and on Mondays and Fridays, 25 people are exercising. How many students in all does he teach each week?

Missing Information: _____

WN 2. The monthly rent for a studio apartment in Mr. Santos's building is $650, plus utilities. A similar apartment in a nearby building rents for $690, including utilities. How much money would you save per month if you rented Mr. Santos's apartment?

Missing Information: _____

P 3. Sandy bought a secondhand drum set for $179.40, including tax. She later bought some sheet music priced at $4.00 plus tax. How much did she pay altogether?

Missing Information: _____

M 4. Mrs. Sobala leaves home at 7:30 A.M. and takes a 25-minute train ride to work. On her way home she takes a bus instead, and that trip takes 35 minutes. What time does Mrs. Sobala get home?

Missing Information: _____

GR 5. Of all the residents in Lane County, how many of them voted for the tax increase?

Missing Information: _____

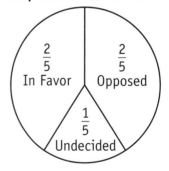

Lane County Residents' Response to Tax Increase

$\frac{2}{5}$ In Favor

$\frac{2}{5}$ Opposed

$\frac{1}{5}$ Undecided

 Select the right answer for each of the following word problems. Some of the problems have enough information for you to solve them, but others do not.

WN 6. Harold started the day with $45.00 in his wallet. If he bought 2 gallons of milk at $3.49 each, then bought 14 gallons of gas, how much money did he have left in his wallet?

    a. $26.57   **d.** $42.91

    b. $28.66   **e.** not enough information given

    c. $40.82

D 7. A pizza parlor spends approximately $4.70 on the ingredients to make one large pizza. In addition, it pays about $2.10 for the labor and all other costs in making one large pizza. If the parlor charges $10.75 for one large pizza, what is its profit?

    a. $3.95   **d.** $11.55
    b. $5.05   **e.** not enough information given
    c. $5.80

P 8. The Guerrero family's lunch bill came to $36.50 before 6% tax was added to the bill. If Mr. Guerrero paid for the lunch with a $50 bill, how much change did he receive?

    a. $3.00   **d.** $26.00
    b. $11.31   **e.** not enough information given
    c. $13.50

M 9. A bus driver left Station A at 4:45 P.M. and drove for 1 hour and 15 minutes at 60 miles per hour. He stopped at Station B to refuel and pick up more passengers. He then drove for another hour at 55 miles per hour to get to Station C. What time did the bus arrive at Station C?

    a. 6:30 P.M.   **d.** 8:30 P.M.
    b. 7:00 P.M.   **e.** not enough information given
    c. 8:00 P.M.

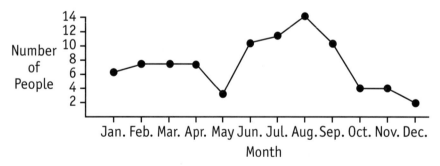

**Mavco Industries'
Layoffs Per Month**

GR 10. According to the graph, how many people were employed by Mavco Industries in June, July, and August?

    a. 35   **d.** 10
    b. 14   **e.** not enough information given
    c. 12

What More Do You Need to Know?

Nancy bought some pants at one store and a winter coat at another. Both were 20% off the original price. Before tax, the sale price of the coat was $140. What more do you need to know to find the total amount that Nancy spent on the coat?

a. the sale price of the pants
b. the original price of the pants
c. the tax on the coat
d. the total paid for pants and coat
e. the original price of the coat

Is there enough information given to find how much Nancy spent on the coat?
What more do you need to know to solve the problem?

Here's another example of a word problem in which there is not enough information to get an answer. This type of problem asks you to identify what piece of additional information is needed.

If the price of the coat is $140 plus tax, then you need to know the amount of tax to find the total.

$$
\begin{array}{l}
\$140 \\
+\ \ ??\ \ \leftarrow \text{tax} \\
\hline
\$\ ???\ \leftarrow \text{total}
\end{array}
$$

The answer to this problem is **c. the tax on the coat.**

Try the next exercise to get some practice in working with this type of word problem.

Decide what more you need to know to solve each problem.

WN 1. Each crate of books that Randall fills in his packing job weighs 76 pounds. He packs a total of 3,800 pounds in one day. What more do you need to know to find out how many books he packs in a day?

    a. the number of pounds he packs in an hour
    b. the weight he packs in 1 hour
    c. the number of books in a crate
    d. the number of hours he works in a day
    e. the number of books he packs in an hour

D 2. Regina paid $2.50 for a hairbrush, $1.76 for a package of combs on sale for 10% off, $0.69 for each of two barrettes, and $0.95 in tax. What more do you need to know to find how much change she received from the cashier?

    a. the total cost of the barrettes
    b. the tax rate in her state
    c. the original price of the combs
    d. the amount of money she gave the cashier
    e. the total number of items she bought

D 3. After driving 70 miles, a driver has about 12 gallons of gas left in his tank. He fills it with gas priced at $1.31 per gallon. What more do you need to know to find how much he paid in all for the gas?

    a. the amount of gas his tank holds
    b. the miles per gallon his car gets
    c. the miles per hour he drove
    d. the price of his last fill-up
    e. the number of hours he drove

P 4. Twenty-three percent of the people who responded to a questionnaire said that they were pleased with the performance of the government. A total of 35,000 questionnaires was sent out. What more do you need to know to find out the number of people who said they were pleased with the government's performance?

    a. the number of people in the country
    b. the percent that said they were displeased
    c. the number of people who responded to the questionnaire
    d. the number of answer choices for the question
    e. the percent of questionnaires sent out

GE 5. The rug shown here is placed lengthwise on a floor that is 15 feet long. What more do you need to know to find the fraction of the floor that the rug covers?

4 ft

8 ft

    a. the area of the rug
    b. the ratio of the length of the rug to its width
    c. the area of the floor covered by furniture
    d. the difference in length between the rug and the floor
    e. the width of the floor

Working with Item Sets

Read the information and questions below. Don't answer them yet.

Gwen and Richard Harris have a daughter named Maggie for whom they need to provide child care. They work an average of 20 days a month.

At Little Angels Center, the monthly charge is $820. Little Angels also charges a one-time $50 registration fee, a one-time yearly $25 insurance charge, and $15 monthly to cover the cost of field trips.

However, Gwen and Richard are also considering having their friend, Mrs. Mitchell, baby-sit in her home for $6 an hour.

1. How much would Gwen and Richard pay for the first month at Little Angels?

2. If Maggie stays at Mrs. Mitchell's for 10 hours a day, what would be the Harris's average monthly child-care cost?

3. After the first month, what would be the cost difference between Mrs. Mitchell's home day care and Little Angels Center?

This type of word problem is called an **item set.** It is made up of one or more short paragraphs followed by two or more questions.

What seems difficult about item sets?
What's the best way to handle them?

Item sets may seem hard because so much information is given. There is a lot of reading to do. In addition, having more than one question per problem may seem confusing.

Don't worry. You can successfully handle item sets by using the same skills you use when you sort necessary and unnecessary information. Here are some tips that can help.

1. **READ ALL THE INFORMATION SLOWLY AND CAREFULLY.**
    Don't rush into the problems. Slowly read the information once or
    twice until you get a "picture" of what is being presented.

2. **ANSWER EACH QUESTION SEPARATELY.** Don't try to read all
    the questions at once. Even though they are listed together as a set,
    the questions are exactly like those of regular word problems if you
    deal with them one at a time. Look at Question 1 from page 71
    again.

    How much would Gwen
    and Richard pay for the
    first month at Little Angels?

    $$\$820 \;+\; \$50 \;+\; \$25 \;+\; \$15 \;=\; \$910$$

    ↑ monthly charge ↑ registration ↑ insurance ↑ field trips

3. **AFTER READING A QUESTION, SKIM THROUGH THE
    PARAGRAPH AND JOT DOWN THE INFORMATION
    NECESSARY FOR JUST THAT QUESTION.** Use Question 2 from
    page 71 to see how you need to skim for the necessary information.

    If Maggie stays at
    Mrs. Mitchell's for
    10 hours a day, what
    would be the Harris's
    average monthly
    child-care cost?

    STEP 1 Find total hours.
    20 days × 10 hours = 200 hours
    ↑ Harris's work schedule ↑ information within Question 2

    STEP 2 Find total costs.
    200 hours × $6 = **$1,200**

4. **BE ESPECIALLY CAREFUL WITH MULTISTEP PROBLEMS.**
    Some problems take two or more steps to solve. Questions 3 from
    page 71 illustrates this.

    After the first month, what
    would be the cost difference
    between Mrs. Mitchell's
    home day care and Little
    Angels Center?

    STEP 1 Little Angels cost
    $820 + $15 = $835
    ↑ monthly charge ↑ field trips

    STEP 2 Mrs. Mitchell cost
    20 days × 10 hours × $6 = $1,200

    STEP 3 Compare.
    $1,200 − $835 = **$365**

Read each item set carefully. For each question, write the information needed to solve only that problem; then solve it.

At the music box factory, each assembly line manufactures 80 boxes per hour when it works at top speed. On Tuesday, eight lines worked the day shift and four lines worked the night shift. Each worker on the day shift earns $6 per hour, while the workers on the night shift earn $9 per hour. There are 4 workers on each assembly line at all times.

WN 1. How many people worked on the assembly line on Tuesday?

Necessary Information: _____

Answer: _____

WN 2. How much did the factory pay in hourly wages on Tuesday if each shift is 8 hours long?

Necessary Information: _____

Answer: _____

WN 3. How many boxes are manufactured in a day shift of 8 hours on Tuesday if the lines are working at top speed?

Necessary Information: _____

Answer: _____

Myra and Henry did some yard work behind their home. First they dug up a concrete area that measured 20 feet by 12 feet. They planted grass in three-fourths of that space, and they plan to put a brick patio in the rest. They will need four bricks per square foot for the patio. If they build the patio themselves, their only cost will be the bricks at $0.20 each and $20 for a bag of mortar. Jack's Landscape Co. wants $320 for the entire job—including labor and materials.

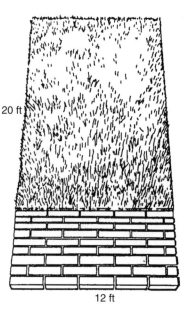

20 ft

12 ft

GE 4. How many square feet are in the grassy area of the yard?

Necessary Information: _____

Answer: _____

GE 5. How many square feet are in the area where the brick patio will be built?

Necessary Information: _____

Answer: _____

GE 6. How many bricks will Myra and Henry need to build their patio?

Necessary Information: _____

Answer: _____

D 7. If Myra and Henry decide to do the patio themselves, how much less will it cost them than if Jack's Landscape Co. does the job?

Necessary Information: _____

Answer: _____

Making Charts to Solve Item Sets

Gloria and Robert Shaw want to buy a new car that has a list price of $18,500. Gloria's monthly take-home pay is $1,800, and Robert's is $2,400. They would have to make a down payment of $1,850 (10%) and monthly payments of $510 for 36 months.

What makes item sets difficult?
What can simplify them for you?

Many times there is so much information in an item set that it is easy to get confused. All of the facts and numbers can make it hard to select the information you need to solve each question. Here's a strategy that can help. Look at how you can organize the information from the item set into a chart. Then you can use the chart to select the information you need.

Income	Car Costs
Gloria = $1,800 per month	• $18,500 list price
	• $1,850 down (10%)
Robert = $2,400 per month	• 36 months of payments at $510 per month

1. How much do Gloria and Robert have to pay off after they make the down payment?

   Information: $18,500; $1,850 down

ANSWER: $18,500 – $1,850 = **$16,650**

2. What total price will the Shaws pay for the car?

   Information: $1,850 down; 36 months at $510 per month

   STEP 1 36 months × $510 = $18,360

   STEP 2 $1,850 down + $18,360 = $20,210

ANSWER: $20,210

3. How much will the Shaws have left per month after they make the car payment?

Information: $1,800; $2,400; $510

> **STEP 1** $1,800 + $2,400 = $4,200
>
> **STEP 2** $4,200 − $510 = $3,690

ANSWER: $3,690 per month

• •

 Read the item set below and fill in the chart provided. Then answer the questions that follow.

Sasha was comparing two projects she had worked on. The first project took her 100 hours to complete. For this project Sasha required 10 hours of help from a computer operator who earns $7 per hour. Expenses for materials came to $124. The second project that Sasha worked on required no outside help, but it took her 20 entire workdays to complete. The materials cost $280. Sasha earns $9 per hour, and her workday consists of 9 hours minus 1 hour for lunch.

	First Project	**Second Project**
Sasha's pay at $9 per hour		
Computer operator's pay at $7 per hour		
Cost of materials		

WN **1.** For materials, how much more did the company have to pay for the second project than the first?

a. $404

b. $260

c. $156

d. $150

e. not enough information given

WN 2. What total dollar amount did the first project cost the company?

    a. $9
    b. $124
    c. $700

    d. $900
    e. $1,094

WN 3. What total dollar amount did the second project cost the company?

    a. $7
    b. $1,440
    c. $1,720

    d. $1,724
    e. not enough information given

For the item set below, fill in the chart to organize the information. Then answer the questions.

Giant Grocery sells carrots at $1.19 per pound or $4.00 for a 5-pound bag. The owner of the store gives a 15% discount on everything *except* vegetables to local restaurants. An Italian restaurant down the street uses about 20 pounds of carrots and 25 pounds of flour each week.

Cost of carrots	
Restaurant purchases	
Discount rates	

D 4. If the Italian restaurant bought bags of carrots to last a week, how much money would it save if it bought carrots in 5-pound bags?

    a. $3.20
    b. $7.80
    c. $9.31

    d. $9.71
    e. not enough information given

P 5. If a nearby restaurant purchased 3 bags of sugar at $3.60 per bag, how much did it pay?

    a. $0.39
    b. $1.62
    c. $9.18

    d. $10.80
    e. not enough information given

D 6. How much does the Italian restaurant pay for the flour it uses each week?

    a. $3.75
    b. $7.50
    c. $8.25

    d. $9.88
    e. not enough information given

Read the item set below and fill in the chart provided. Then answer the questions that follow.

Sue Maynard bought $5\frac{1}{2}$ yards of cotton fabric to make three baby outfits. One outfit called for 2 yards; the other two called for $\frac{3}{4}$ yard each. The fabric costs $4.00 per yard plus 5% tax.

Yards needed			
Cost per yard			
Tax			

F **7.** How many yards of fabric are needed for the three outfits?

    a. 9   **d.** 2

    b. $5\frac{1}{2}$ **e.** $1\frac{1}{2}$

    c. $3\frac{1}{2}$

F **8.** How many yards of fabric did Sue have left after she made the three outfits?

    a. 9   **d.** 2

    b. $5\frac{1}{2}$ **e.** 0

    c. $3\frac{1}{2}$

P **9.** How much did Sue spend on the fabric she bought?

    a. $0.70   **d.** $23.10

    b. $1.10   **e.** $23.80

    c. $14.70

D **10.** If Sue gave the cashier $30.00, how much change did she receive?

    a. $6.20   **d.** $25.90

    b. $6.90   **e.** $29.30

    c. $15.30

Mixed Review

 Choose the correct answers for the following word problems.

WN 1. The Hair Etc. chain of hair salons gave 340 haircuts and 210 permanents in July. It gave 420 haircuts and 185 permanents in August. How many more haircuts were given in August than July?

    a. 25   **d.** 365
    b. 80   **e.** 410
    c. 130

WN 2. To settle an argument among his three children, Bob took the 12 marbles Anna had, the 27 marbles Bea had, and the 33 marbles Gary had and divided them equally among the children. How many marbles did each child get?

    a. 216   **d.** 24
    b. 72   **e.** 8
    c. 27

D 3. Before a rat was injected with Drug X, it traveled through a maze in 12.1 seconds. After the injection the same rat traveled the maze in 15.3 seconds. A second rat, with no injection, completed the maze in 15.2 seconds. How many seconds slower was the first rat after the injection?

    a. 3.2   **d.** 0.32
    b. 3.1   **e.** 0.31
    c. 3.0

F 4.

Day	Adult	Child
Friday	870	435
Saturday	1,115	224
Sunday	1,002	331

The chart above shows the attendance numbers for the basketball playoffs at Milton High. Adult tickets sold for $3.50; children's tickets sold for half that price. How much money was collected from children's tickets for all 3 days?

    a. $5,227.25   **d.** $175.00
    b. $3,465.00   **e.** not enough information given
    c. $1,732.50

F 5. One week Al's Advertising delivered $\frac{1}{4}$ of its fliers on Monday, $\frac{1}{5}$ of its fliers on Friday, $\frac{2}{5}$ of its fliers on Saturday, and the rest on Sunday. What fraction of its fliers did Al's deliver on the weekend?

a. $\frac{1}{20}$ d. $\frac{2}{5}$

b. $\frac{3}{20}$ e. $\frac{11}{20}$

c. $\frac{1}{5}$

M 6. On one workday Jill had three different meetings. One lasted only 15 minutes, another was an hour and 15 minutes long, and the last one was 55 minutes long. Which of the following expressions show the number of *hours* Jill spent in meetings?

a. $15 + 1.15 + 55$ d. $\dfrac{15+75+55}{60}$

b. $\dfrac{15+1.15+55}{60}$ e. $\dfrac{15+15+55}{60}$

c. $(15 + 75 + 55) \times 60$

Problems 7–8 refer to the following passage.

Ed's television is not working. He is deciding whether to have it repaired or to buy a new set. He can have it fixed for $150. A new television at Bert's Discount will cost him $305, plus an extra $75 for a warranty. The same set at Tech TV will cost him $400, including the warranty.

WN 7. How much more will the television plus warranty cost at Tech TV than at Bert's?

a. $20   d. $155

b. $80   e. not enough information given

c. $95

WN 8. Which expression shows the amount of money Ed will save by getting his television repaired instead of buying one at Bert's without a warranty?

a. $305 – $150 – $75   d. $400 – $150

b. $305 + $75 – $150   e. $305 + $150

c. $305 – $150

MAKING A PLAN

Choosing the Operation

A carpenter needs 24 nails to put together 1 shelving unit. She wants to build 6 units. How many nails in all will she need?

a. 4   **d.** 144
b. 18   **e.** not enough information given
c. 30

What is the correct answer?
How did you decide whether to add, subtract, multiply, or divide?

In the problem above, you are given a group (24 nails), and you are asked to find the total in 6 groups. Therefore, you multiply 24 by 6 to find that the answer is **d. 144.**

Remember that when you are trying to decide how to solve a word problem, you should picture the problem to understand what is going on. The chart on the next page shows how picturing the problem can help you choose the right operation.

If the problem gives	and you want to	you should	
A two or more numbers	put them together in the same group	add	
B one number being taken away from another	find out what is left	subtract	
C two amounts	find out how much larger one is than the other	subtract	
D several items in a group and a number of groups	find the *total* number in all of the groups	multiply	
E the cost, weight, or measure of *one* item	find the *total* cost, weight, or measure of many	multiply	
F the total number of items and a number of groups	find the number of individual *items* in each group	divide	
G a total number of items and the number in each group	find the number of *groups*	divide	

TIP
Notice the answer choices in the problem on page 81. In many multiple-choice questions, some of the incorrect answer choices are based on using the wrong operation. (For example, **a. 4** is based on $24 \div 6 = 4$.)

Using the ideas on page 82, decide which operation you should use to solve each problem. Write the symbol of the operation in the brackets. [+, −, ×, or ÷] Then solve the problem using the correct operation.

EXAMPLE   Betty ran 5 miles last Thursday. On Tuesday she ran 7 miles. How many more miles did she run on Tuesday than on Thursday?

Operation [−]   Answer: *7 − 5 = 2 miles*

WN 1. When his class took a field trip yesterday, Mr. Rodríguez divided the students into groups of 8. The total class size was 40 students. How many groups were there?

Operation [ ]   Answer: _____

WN 2. The Fontines had 655 bricks to sell. One customer came and purchased 435 of them. How many bricks were left?

Operation [ ]   Answer: _____

WN 3. There were 96 children, 45 men, and 20 women at the festival on Saturday. How many people were there in all?

Operation [ ]   Answer: _____

WN 4. Ted used 4 dozen eggs in his cooking for the senior citizens' home. How many single eggs did he use?

Operation [ ]   Answer: _____

WN 5. How many ounces larger is Can A than Can B shown at the right?

Operation [ ]   Answer: _____

A   B
15 ounces   9 ounces

D **6.** Catherine spent $54.90 at the grocery store last week. This week she spent $50.75. How much more did Catherine spend last week?

Operation [   ]   **Answer:** _____

D **7.** A dry cleaner charges $1.10 per shirt for overnight service. If Terry drops off 5 shirts, how much will he have to pay?

Operation [   ]   **Answer:** _____

D **8.** The supervisor at Pente Construction paid out a total of $750.50 in wages today. Each worker received $150.10. How many workers were paid?

Operation [   ]   **Answer:** _____

M **9.** Sam is 4 feet $9\frac{1}{2}$ inches tall and his older brother Jack is 5 feet $6\frac{1}{2}$ inches tall. How much taller is Jack?

Operation [   ]   **Answer:** _____

M **10.** Tyrone cut the board below into four equal pieces. How many centimeters was each piece?

104 cm

Operation [   ]   **Answer:** _____

Equations

To qualify for the swim team, Jeanette needs to swim a total of 325 laps. Jeanette will swim the same number of laps every day over the next 5 days. How many laps will she need to swim daily to qualify?

How do you decide what operation to use to solve this problem? Try rewording the problem <u>using no numbers.</u> Then write a plan to help you solve it.

Here is an example of the problem written without numbers. Decide on a plan for solving it, then write an **equation** (also called a **number sentence**) to organize your plan.

REWORD: Jeanette needs to swim some laps. She wants to swim the same number of laps for several days. How many laps does she need to swim each day?

PLAN: Divide total laps by the number of days to find the number of laps each day.

EQUATION: $\underset{\text{total laps}}{325} \quad \underset{\text{operation}}{\div} \quad \underset{\text{number of days}}{5} \quad \underset{\text{equals}}{=} \quad \underset{\text{laps per day}}{}$

Jeanette will need to swim 325 ÷ 5 or **65 laps** each day.

Here's another example.

The original price of a handmade scarf was $20.00. The designer, however, had to raise the price to $23.50 to cover some unexpected expenses. By how much was the price of the scarf raised?

REWORD: The original price of a scarf was raised. By how much was the price raised?

PLAN: Take the new price and subtract the old price to find the amount the price was raised.

EQUATION: $\underset{\text{new price}}{\$23.50} \quad \underset{\text{operation}}{-} \quad \underset{\text{old price}}{\$20.00} \quad \underset{\text{equals}}{=} \quad \underset{\text{amount raised}}{\$3.50}$

> **TIP**
> When you are writing an equation, write the labels with the numbers. This will help you picture the problem and see if your answer makes sense.

Choose the correct equation for each of the following problems.

WN **1.** Mara babysat for 5 hours on Monday. For two of those hours the children were sleeping. How many hours were the children awake?

    a. $5 + 2 = 7$ hours   **b.** $5 - 2 = 3$ hours

WN **2.** Peter has 117 more miles to go on a 220-mile trip. How many miles has he traveled so far?

    a. $220 - 117 = 103$ miles   **b.** $220 + 117 = 337$ miles

WN **3.** Tremaine collected 140 cans on Saturday morning. His sister collected 30 cans more than Tremaine. How many cans did Tremaine's sister collect?

    a. $140 - 30 = 110$ cans   **b.** $140 + 30 = 170$ cans

D **4.** A farmer sells eggs for $2.09 per dozen. In a normal week he sells 35 dozen eggs. How much money does he earn each week from egg sales?

    a. $\$2.09 \times 35 = \$73.15$ **b.** $\$2.09 \div 35 = \$0.06$

F **5.** The clerk at Hannah's Fabric Store cut this piece of cotton into strips that measured $\frac{1}{3}$ foot wide. How many strips did she get out of the cotton?

    a. $6 \div \frac{1}{3} = 18$ strips   **b.** $6 \times \frac{1}{3} = 2$ strips

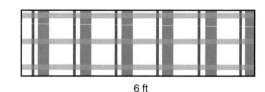

6 ft

Write a plan and an equation for each of the following problems. Then solve the problem.

WN **6.** Carl works 8 hours each day. If his workweek is 6 days, how many hours does Carl work per week?

PLAN:

EQUATION:

WN **7.** An average load of laundry at the CleanMart takes 17 minutes to wash and 45 minutes to dry. How long does it take to wash and dry one load?

PLAN:

EQUATION:

WN **8.** Frances drove 76 miles of the 300-mile distance between her home and her ex-husband's. How many more miles does she need to travel?

PLAN:

EQUATION:

WN **9.** According to the sign at the right, if a customer had 6 shirts dry-cleaned, how much would he or she pay?

PLAN:

EQUATION:

Dry-Cleaning Prices		
Suits	→	$9.00
Shirts	→	$1.00
Pants	→	$3.50
Coats	→	$11.00

WN **10.** The people attending a nurses' conference were asked to break up into small groups of 15 people. There were 225 people in all at the conference. How many small groups were made?

PLAN:

EQUATION:

WN 11. Two families that live on the same block had a yard sale together. One family took in $87 for the day; the other family sold $112 worth of items. How much money did the families earn in all?

PLAN:

EQUATION:

WN 12. A landscaper planted a new tree every 8 feet along one side of this path. How many trees did she plant?

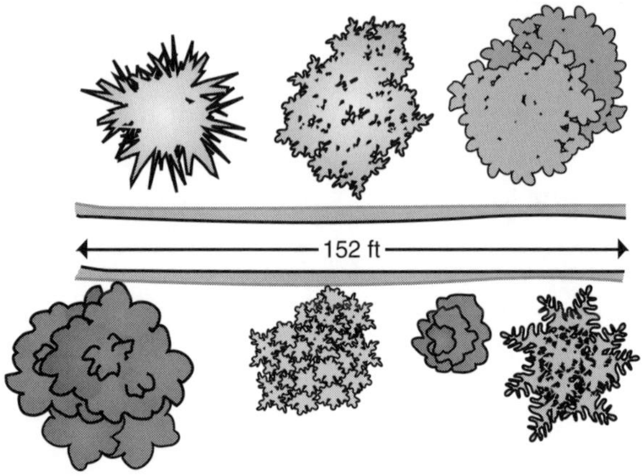

PLAN:

EQUATION:

152 ft

F 13. Mary's sandbox holds 3 cubic feet of sand. Right now it is only $\frac{1}{2}$ full. How many cubic feet of sand are in the box?

PLAN:

EQUATION:

P 14. Fourteen percent of the students at North High are in the vocational program. If there are 800 students at the school, how many are in the program?

PLAN:

EQUATION:

GE 15. How many inches of braid will a tailor need to trim all sides of the square shown at the right?

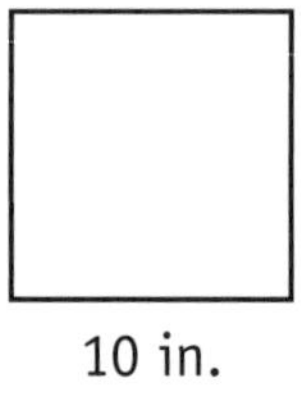

PLAN:

EQUATION:

10 in.

Equations with Two Operations

The Elderberry Taxi Service charges $2 for a trip into town and $1 for the return trip. Isabel took two round-trips into town this past week. How much did she pay in all?

How many operations do you need to solve the problem?

Your plan for this problem will involve more than one operation. Perhaps you would think something like this:

PLAN: First I should add the costs of the trips to and from town. Then I should multiply the total cost by the number of round-trips.

Using what you learned about set-ups, you could write

$$(\$2 + \$1) \times 2 = \textbf{\$6}$$

costs   trips

This can also be written as $2(\$2 + \$1) = 6$.

As you learned, you should do the operation **inside the parentheses first.** In this case, the parentheses tell you to add first, then multiply.

Let's look at another problem with more than one operation.

Ramón started the day with $35.70 in his wallet. He spent $4.50 on breakfast and bus fare, then divided the remaining money equally between his wife and his daughter. How much money did Ramón's wife receive?

PLAN: I should subtract the money Ramón spent from the amount he started with. Then I should divide the remaining amount by the number of people he gave it to.

$$(\$35.70 - \$4.50) \div 2 = \textbf{\$15.60}$$

money he   amount   wife &
started with   spent   daughter

Again, see that the parentheses in the equation tell you which operation should be performed first. Look at what would happen if you did the dividing *before* the adding.

RIGHT: ($35.70 − $4.50) ÷ 2   WRONG: $35.70 − ($4.50 ÷ 2)

    = $31.20 ÷ 2   = $35.70 − $2.25

    = $15.60   = $33.45

> **Remember This**
> **Order of Operations:**
> 1. Solve within parentheses first.
> 2. Do multiplication and division next, from left to right.
> 3. Do addition and subtraction last, from left to right.

When you read a word problem, decide whether it will take one operation or more than one operation to solve. Then write an equation that shows all of the operations.

....

Choose the correct equation that represents each problem below.

WN 1. Mary has taken 18 sick days in 3 years. On average, how many sick days has she taken each month?

    a. $18 \div (3 \div 12) = 72$ **b.** $(18 \div 3) \div 12 = \frac{1}{2}$

WN 2. An art teacher divided a box of 36 crayons among 4 students. He then added 3 crayons to each student's pile. How many crayons did each student get?

    a. $(36 \div 4) + 3 = 12$ **b.** $36 \div (4 + 3) = 5.14$

D 3. Inés's electric bill came to $64.89 last month. A total of $14.01 of that amount was for services and tax; the remaining amount was for the kilowatt-hours of energy she consumed. If the electric company charges $0.06 per kilowatt hour, how many kilowatt hours did Inés consume?

    a. $\$64.89 − (\$14.01 − \$0.06) = 50.94$
    b. $(\$64.89 − \$14.01) \div \$0.06 = 848$

D 4. A building company is deciding whether or not to purchase two adjacent lots; one lot measures 1.75 acres, and the other measures 2.8 acres. The owner will sell the land for $1,244 per acre. How much would both lots cost?

    a. $(1.75 + 2.8) \times \$1,244 = \$5,660.20$
    b. $1.75 + (2.8 \times \$1,244) = \$3,484.95$

M **5.** A factory worker is responsible for cutting plastic tubes like the one shown into 2-inch lengths. If the worker cuts 20 tubes in 1 hour, how many 2-inch lengths will he have at the end of an hour?

2 ft

    a. $(24 \times 20) \div 2 = 240$ **b.** $24 \div (20 \div 2) = 2.4$

 Write an equation for each of the following problems. Then solve the problems.

WN **6.** Lisa made 3 dozen cupcakes to take to her office on Valentine's Day. She left 14 cupcakes on the second floor, then divided the remaining cupcakes among the 22 people who work on the fifth floor. How many cupcakes did each worker on the fifth floor receive?

    EQUATION:

WN **7.** Computer World's newest printer can print an average of 400 words per minute. How long would it take for the printer to print Ms. Savage's weekly report if that report is 15 pages long with 200 words per page?

    EQUATION:

D **8.** Mrs. Teel bought two swordfish at the local market; one weighed 3.5 pounds, and the other weighed 2.0 pounds. If the cost of the fish totaled $45.65 before tax, how much did they cost per pound?

    EQUATION:

F **9.** Last week Howard worked $8\frac{1}{2}$ hours on Monday, $4\frac{1}{2}$ hours on Tuesday, and $6\frac{1}{2}$ hours on Friday. If Howard earns $7.10 per hour, how much did he earn that week?

    EQUATION:

F 10. Annie Gordon swam $5\frac{1}{2}$ miles this week. Her brother Paul swam $8\frac{1}{2}$ miles. What is the average number of miles the two swam?

EQUATION:

P 11. Twenty percent of the pounds lost by the Winners' Weight Loss group were lost during the first month of the program. According to the chart, how many pounds were lost in that month?

EQUATION:

Winners' Total Weight Loss	
Member	**Pounds**
Kay Hughes	4
Isabel Perkins	8
Janet Liao	5
Margaret Oates	3
Frank Rodríguez	5

GE 12. Juan's vegetable garden measures 8 feet by 7 feet. The mulch he plans to buy covers 28 square feet per bag. How many bags does he need to cover the garden?

EQUATION:

D 13. Maria took her friend Julia to lunch. Julia's lunch cost $5.95 and Maria's cost $6.50. They also shared a pitcher of cola for $2.75. How much change should Maria receive from $30.00 when she pays for their lunches?

EQUATION:

Writing Equations for Word Problems

Ted had $35 in his wallet one morning. At his office, he collected some money from a bet he had made. When he got home, he counted $73 in his wallet. How much money did Ted collect at the office?

What operation should you use to solve this problem? How did you decide?

Here is a fact about word problems that you may not have thought of.

IN EVERY WORD PROBLEM SOMETHING *IS EQUAL TO* SOMETHING ELSE.

In other words, all word problems give you certain numbers and tell you that these numbers **are equal to** other numbers. Knowing this fact will help you write **equations** to solve all kinds of word problems.

$35	+	■	=	$73
↑	↑	↑	↑	↑
money Ted started with	added to	amount he got at office	equals	$73

You could represent the equation in several ways.

$$\$35 + ■ = \$73$$

- The box ■ stands for the amount Ted collected at work. You could also write the equation

$$\$35 + x = \$73$$

- The letter x is often used in equations to stand for *the unknown*—what you are being asked to find.

When you write an equation, you are representing both sides of the equals sign (=) as two equal amounts. Think of this as balancing two sides of a scale.

In other words, what would you have to add to $35 to "balance" the $73 on the other side? You could decide that the answer was **$38.** The next lesson will give you some techniques for solving equations.

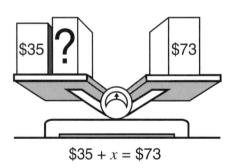

$$\$35 + x = \$73$$

Now look at another word problem and equation.

Ann Lih baked some cookies and distributed them to 4 nursing homes in her neighborhood. Each nursing home received 60 cookies. How many cookies did Ann Lih bake in all?

Let *x* stand for what you do not know—the total number of cookies baked.
What equation could you write for this problem?

$$\underline{\quad x \quad} \quad \underline{\quad \div \quad} \quad \underline{\quad 4 \quad} \quad \underline{\quad = \quad} \quad \underline{\quad 60 \quad}$$

total cookies   divided by   number of   equals   cookies
                         homes   per home

$$x \div 4 = 60$$

> **TIP**
> Your first step when writing an equation is to make *x* stand for the number you do **not** know in the problem.

What other equations could you write for this problem?

Perhaps you knew from experience that you could multiply to find out what *x* is.   OR   You may have chosen to divide the total cookies by the number of cookies for each home.

cookies   total
per home   cookies
↓   ↓
$$60 \times 4 = x$$
↑
number of homes

cookies per
home
↓
$$x \div 60 = 4$$
↑   ↑
total   number of
cookies   homes

Each of these equations can be used to find the answer to the word problem. In the next exercise, you'll get some practice in writing different equations based on word problems.

For each problem decide what *x* should stand for—the unknown number you will need to find. Then fill in the blanks of each equation, using *x*. Do not solve the equations.

EXAMPLE   A total of 17 people showed up for the first day of a computer class. Several of them arrived late. If 11 people came on time, how many were late?

Let x stand for: *number who were late*

$17 - \underline{x} = 11$ OR $17 - 11 = x$

WN 1. So far this week, Enid has worked 28 hours. How many more hours must she work before she has put in her required 40 hours?

Let *x* stand for: _____

_____ $+ x =$ _____   OR   _____ $- 28 =$ _____

D 2. The cost of a phone call Sally made to Doug was $0.90 per minute. The total bill for the call was $5.40. For how many minutes did Sally and Doug talk?

Let *x* stand for: _____

$\$0.90 \times$ _____ $=$ _____   OR   _____ $\div \$0.90 =$ _____

D 3. Tania recently spent a total of $44.99 on denim pants for her children. She bought three pairs of pants—all at the same price. What was the cost of a single pair of pants?

Let *x* stand for: _____

$3 \times$ _____ $=$ _____   OR $\$44.99 \div$ _____ $=$ _____

GE 4. Mrs. Murray plans to use all of a 120-square-foot piece of carpet to cover her laundry room floor. The length of the room is 12 feet. How wide is the room if the piece of carpet is big enough to cover the whole room?

Let *x* stand for: _____

$12 \times$ _____ $=$ _____   OR $120 \div$ _____ $=$ _____

For each problem first decide what *x* will stand for. Then write two different equations using *x*. Do not solve the equations.

EXAMPLE   Patrick delivered 87 of his newspapers by car. To deliver the remaining 21 papers, he used a wagon. How many newspapers did Patrick deliver?

Let *x* stand for: *total papers delivered*

Equation #1: $x - 87 = 21$ Equation #2: $21 + 87 = x$

WN 5. An employee of the city subway worked a 35-hour week. If he worked the same number of hours on each of 7 days, how many hours did he work per day?

Let *x* stand for: _____

Equation #1: _____   Equation #2: _____

WN 6. Many of the people in Dora's women's group live outside the city. Of the 29 people in the group, only 7 live in the city. How many people in Dora's group are from outside the city?

Let *x* stand for: _____

Equation #1: _____   Equation #2: _____

WN 7. Gary had 80 bunches of daisies to sell to passing cars. The total number of daisies in his cart was 720. How many daisies made up each bunch?

Let *x* stand for: _____

Equation #1: _____   Equation #2: _____

WN 8. Vanessa has $21 to spend on party favors for her son's birthday. Seven boys will be at the party. How much money can Vanessa spend on each boy?

Let *x* stand for: _____

Equation #1: _____   Equation #2: _____

Solving a Word Problem Equation

A butcher cut up several chickens and divided them into 4-pound packages. He ended up with 9 packages. How many pounds of chicken did the butcher start with?

Write an equation for this word problem.
What is the answer to the problem?

Suppose you are not sure how to solve this problem. You write an equation to help you solve it. You let *x* stand for the total pounds of chicken because that is the number you do not know. Your equation would look like this.

$$
\underset{\substack{\uparrow \\ \text{total} \\ \text{pounds}}}{x} \; \div \; \underset{\substack{\uparrow \\ \text{pounds} \\ \text{per pkg}}}{4} \; = \; \overset{\substack{\text{number of pkgs} \\ \downarrow}}{9}
$$

The following guidelines will help you learn to solve equations such as this one.

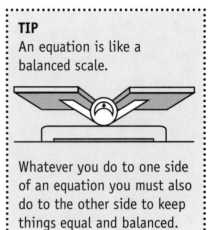

TIP
An equation is like a balanced scale.

Whatever you do to one side of an equation you must also do to the other side to keep things equal and balanced.

1. **TO SOLVE ANY EQUATION, YOU MUST GET THE UNKNOWN (*x*) ALONE ON ONE SIDE OF THE EQUALS SIGN.**

   In the equation, you want *x* to stand alone. In other words, you want to get rid of the ÷ 4 that is next to it.

2. **USE THE OPPOSITE OPERATION TO GET THE UNKNOWN ALONE.**

   To get rid of the ÷ 4, you **multiply** both sides of the equation by 4. Multiplication is the opposite of division.

   $$x \div 4 = 9$$
   $$x \div 4 \times 4 = 9 \times 4$$

3. **SIMPLIFY THE EQUATION.**

   When you divide by 4 and then multiply by 4, the operations cancel each other out and your unknown stands alone.

   $$x \div 4 \times 4 = 9 \times 4$$
   $$x = 9 \times 4$$
   $$x = 36$$

ANSWER: The butcher started with **36 pounds** of chicken.

Take a look at how to solve another word problem using an equation.

Harold deposited some money in his account yesterday, making a balance of $434.90. If his balance was $325.10 before the deposit, how much did Harold put in his account?

**Write an equation based on this problem.
Think about how you would solve it.**

Although there are several different ways to solve this problem, try using this equation.

amount before   total after
deposit   deposit
↓   ↓

$$\$325.10 \; + \; x \; = \; \$434.90$$

↑
amount of
deposit

> **Do You Know
> Math Operation
> Opposites?**
> - The opposite of adding is subtracting.
> - The opposite of subtracting is adding.
> - The opposite of multiplying is dividing.
> - The opposite of dividing is multiplying.

STEP 1   Decide what number needs to be removed to get the unknown alone on one side of the equals (=) sign.

To get x alone in this problem, I need to get rid of the $325.10.

STEP 2   Determine what operation is being performed.

The $325.10 is being added to x.

STEP 3   Perform the **opposite** operation to BOTH sides of the equation.

The opposite of adding is subtracting.
$$\$325.10 - \$325.10 + x = \$434.90 - \$325.10$$

STEP 4   This opposite operation cancels out the numbers next to the unknown. You are left with a new equation with x alone.

$$x = \$434.90 - \$325.10$$
$$x = \bm{\$109.80}$$

"Switching Places" in Equations

Let's take a look at one more example of equations with word problems. This one is a little tricky.

The Sidfield Plant employed 1,300 people in 1998. Due to a slow economy, many people were laid off in early 1999, leaving Sidfield with only 590 workers. How many people were laid off?

Write an equation to solve this problem.
Can you think of more than one equation?

Here is one equation you could write for this problem.

$$\underset{\downarrow}{\overset{\text{1998 total}}{}} \quad \underset{\downarrow}{\overset{\text{1999 total}}{}}$$
$$1,300 \;-\; \underset{\uparrow}{x} \;=\; 590$$
$$\text{number laid off}$$

Because there is a minus sign (–) in front of the unknown, this is a more difficult equation to solve. You would need to perform several more steps to get the x by itself. Now look at another equation for the same problem.

$$\underset{\downarrow}{\overset{\text{1998 total}}{}} \qquad \underset{\downarrow}{\overset{\text{number laid off}}{}}$$
$$1,300 \;-\; \underset{\uparrow}{590} \;=\; x$$
$$\text{1999 total}$$

The two equations—$1,300 - x = 590$ and $1,300 - 590 = x$—have the same solution: **710.** The second one is easier to solve because the x is already alone on one side of the equals sign.

If an equation has an unknown with either a minus (–) or a division (÷) sign in front of it, rewrite the equation as shown below.

switch

EXAMPLE $36 \div x = 9$ becomes $36 \div 9 = x$

$36 - x = 4$ becomes $36 - 4 = x$

switch

Use the steps for solving equations to find x.

1. $37 + x = 91$

2. $x \div 13 = 11$

3. $144 \div 4 = x$

4. $x - 194 = 19$

5. $80 - x = 14$

6. $14 \times x = 42$

7. $144 \div x = 12$

8. $27 + x = 120$

9. $x \div 12 = 9$

10. $105 \div x = 7$

 For each word problem, write an equation and solve it.

EXAMPLE Vera paid for her groceries with a $20 bill. She received $3.14 in change. How much did Vera's groceries cost?

Equation: $20 - x = \$3.14$ OR $20 - \$3.14 = x$

Solve: $20 - \$3.14 = \$16.86$

WN 11. When they drew up their will, Nathan and his wife decided to divide their collection of 96 rare coins equally among their children. If each child is to receive 32 coins, how many children do Nathan and his wife have?

WN 12. A stockperson added 12 pounds of food to a crate, making a total of 60 pounds of food. How many pounds of food were in the crate before the 12 pounds were added?

WN 13. Several committees were formed to plan for the March Against Drugs. Each committee had 15 people. If a total of 225 people were in committees, how many committees were formed?

WN 14. Juan withdrew $145 from his savings account, leaving a balance of $1,004. How much money was in Juan's account before the withdrawal?

WN 15. Hal needed a total of 360 brownies to serve at the club banquet. Each pan in the kitchen held 40 brownies. How many pans did Hal need to fill?

Writing Proportions for Word Problems

In a quality-control check, Rita found that for every 8 usable cups the molding machine was producing 3 rejects. If in 1 hour the machine produced 135 rejects, how many usable cups did it produce?

Can you write an equation to solve this problem?

You learned that in every word problem something *is equal to* something else. But in this problem the equality is hard to see at first. What is equal to what?

In this problem two things are being *compared.* The problem is comparing good cups to rejected cups. One way to compare is to show a **ratio.**

$$\frac{3}{8} \quad \leftarrow \quad \text{rejects}$$
$$\phantom{\frac{3}{8}} \quad \leftarrow \quad \text{usable cups}$$

The ratio is 3 rejects for every 8 usable cups.

The ratio could be written as "3 to 8" or 3:8 or $\frac{3}{8}$. To use ratios to solve word problems, you will be using the fraction form $\frac{3}{8}$.

You should remember that to write an equation for any word problem your first step is always to decide what x, the unknown, will stand for.

What are you being asked to find in this problem?
Let x = _____.

You're right if you let x stand for the **number of usable cups.**

To write an equation, you need to set up a proportion. A **proportion** is made up of two equal ratios.

$$\text{rejects} \rightarrow \frac{3}{8} = \frac{135}{x} \leftarrow \text{rejects} \atop \text{usable cups}$$

rejects →
usable cups →

135 ← rejects
x ← usable cups

Notice that in a proportion the same category, or label, goes on the top of each ratio (rejects), and the same category goes on the bottom (usable cups).

THIS

$$\frac{3}{8} = \frac{135}{x}$$

is NOT the same as

THIS

$$\frac{3}{8} = \frac{x}{135}$$

Later you'll learn how to solve proportion equations. For now, use the next exercise to practice setting up proportions.

> **TIP**
>
> One of the most important steps in writing proportions is remembering to write the numbers in the right places. It is a good idea to first write the ratio in *word* form.
>
> Ratio in word form: $\dfrac{\text{rejects}}{\text{usable cups}}$

Read each of the following problems. Then fill in the missing words and numbers in the proportion. Use x to stand for the unknown number. Do not solve the problems.

EXAMPLE   At Solly's Fresh Produce, bruised apples are sold for $0.90 per dozen. At this price, how much will 26 apples cost?

Compare: $\dfrac{\text{apples}}{\text{cost}}$ $\dfrac{12}{\$0.90} = \dfrac{26}{x}$

WN 1. The Beautiful Bus tour covered 900 miles in 16 hours. If it continues at the same rate, how many hours will it take to cover the remaining 225 miles?

Compare: $\dfrac{\text{miles}}{\text{label} \rightarrow}$ $\quad \dfrac{900}{16} = \dfrac{\quad}{\quad}$ $\quad \begin{matrix} \leftarrow & \text{miles} \\ \leftarrow & \text{unknown} \end{matrix}$

WN 2. The scale of miles on a road map shows that 1 inch represents 300 miles. Todd estimated the road-map distance from Lawton to Dawson to be about 3 inches. How many miles is it from Lawton to Dawson?

Compare: $\dfrac{\text{inches}}{\quad}$ $\quad \dfrac{\quad}{\quad} = \dfrac{3}{x}$

WN 3. Mariel's cake recipe calls for 7 ounces of chocolate to every 3 ounces of sugar. If she uses 21 ounces of chocolate, how many ounces of sugar will she need?

Compare: $\dfrac{\quad}{\text{sugar}}$ $\quad \dfrac{\quad}{3} = \dfrac{\quad}{x}$

D 4. If a 20-ounce package of chicken costs $4.39, how many ounces is a package that costs $5.28?

Compare: $\dfrac{\quad}{\text{cost}}$ $\quad \dfrac{20}{\quad} = \dfrac{\quad}{\$5.28}$

D 5. Apples are advertised $2.95 for a dozen. How much will Mario pay for 8 apples?

Compare: $\dfrac{\quad}{\text{cost}}$ $\quad \dfrac{12}{\quad} = \dfrac{\quad}{x}$

For each of the following word problems, first circle the two things that are being compared. Then set up a proportion using *x* for the unknown number. Do not solve the problems.

EXAMPLE   For every 8 (boxes of greeting cards) that Melanie sells she receives 1 free (movie pass.) How many boxes of cards will Melanie need to sell to earn 4 movie passes?

$$\frac{\text{boxes of cards}}{\text{movie passes}} \quad \frac{8}{1} = \frac{x}{4}$$

WN   6. At the Pasta House, a strand of spaghetti must be 7 inches long to be packaged and sold. On a good day, the workers produce 15 usable strands for every 2 strands that can't be sold. If in one day they had to throw out 1,000 rejects, how many good strands could they package and sell?

WN   7. Mrs. Gutiérrez filled her gas tank and drove 352 miles on 11 gallons. At this rate, how many more miles can she drive on the remaining 4 gallons in her tank?

F   8. To get the right color of paint for Tanisha's apartment, the landlord should mix 2 gallons of blue paint with every 3 gallons of white paint. If the landlord used a total of $5\frac{1}{2}$ gallons of blue paint, how many gallons of white paint did he use?

M   9. A pump can drain 4,800 liters of water from a pond in 50 minutes. How many liters can it drain in an hour and a half?

GE   10. Nancy took a 3-inch-wide by 5-inch-long photograph to a specialty camera shop. She wanted it enlarged to fit exactly in a 20-inch-long frame. How wide will the new picture be?

Solving Proportions

On his test, Willie was told that he answered 9 questions correctly for every 2 questions he got wrong. When he looked at his paper, Willie saw that he got 45 questions correct. How many questions did he get wrong?

Set up a proportion for this word problem.
How can you solve this proportion?

You may have written a proportion like this.

$$\frac{\text{correct}}{\text{wrong}} \quad \begin{array}{c} \rightarrow \\ \rightarrow \end{array} \quad \frac{9}{2} = \frac{45}{x}$$

Or perhaps you chose to put the number of wrong answers on top.

$$\frac{\text{wrong}}{\text{correct}} \quad \begin{array}{c} \rightarrow \\ \rightarrow \end{array} \quad \frac{2}{9} = \frac{x}{45}$$

It doesn't matter which way you choose, as long as **the same categories are on top and the same categories are on bottom.**

> To solve a proportion, remember:
> The **cross products** in a proportion are equal.

For example,

$$\frac{9}{2} \quad \diagdown\!\!\!\!\diagup \quad \frac{45}{x}$$

Cross products are $9 \times x$ and 2×45.

STEP 1 Multiply the cross products.

$$9 \times x = 2 \times 45$$
$$9x = 90$$

STEP 2 Divide to get the x alone.

$$x = 90 \div 9$$
$$x = 10$$

ANSWER: Willie got **10 questions** wrong on his test.

Solve the next problem by following the steps.

A car at Hughes Motor Sales cost $7,200 with a tax of $240. At this same rate, how much tax would a customer pay on a car that costs $9,000?

STEP 1 Decide what numbers are being compared.

The problem is comparing car cost to tax.

STEP 2 Decide what x will stand for.

The unknown is the tax on a $9,000 car.

STEP 3 Set up a proportion with the same categories on top.

$$\frac{tax}{cost} \qquad \frac{240}{7{,}200} = \frac{x}{9{,}000}$$

STEP 4 Write an equation using cross products.

$$240 \times 9{,}000 = 7{,}2000 \times x$$

STEP 5 Solve the equation.

$$2{,}160{,}000 = 7{,}200 \times x$$
$$2{,}160{,}000 \div 7{,}200 = x$$
$$300 = x$$

ANSWER: The tax on a $9,000 car would be **$300**.

....

Cross multiply and divide to solve the following proportions.

1. $\dfrac{x}{5} = \dfrac{4}{10}$

2. $\dfrac{8}{4} = \dfrac{60}{x}$

3. $\dfrac{x}{16} = \dfrac{200}{80}$

4. $\dfrac{9}{x} = \dfrac{6}{4}$

5. $\dfrac{13}{5} = \dfrac{x}{10}$

6. $\dfrac{8}{3} = \dfrac{24}{x}$

7. $\dfrac{x}{9} = \dfrac{25}{45}$

8. $\dfrac{14}{x} = \dfrac{7}{4}$

9. $\dfrac{14}{8} = \dfrac{x}{40}$

 Go back to the proportions you wrote on page 104. Solve for x.

When Can You Use a Proportion?

The assembly line on which Misha works can wrap 93 boxes of pencils in 2 hours. How many hours will it take for this line to wrap an order of 465 boxes?

Can you use a proportion to solve this problem? How do you know?

When you are deciding how to solve a word problem, use this test to find out whether or not you can use a proportion.

boxes and hours

STEP 1   Find two things in the problem that are being compared.

STEP 2   Make a chart and fill in the two things being compared.

boxes		
hours		

STEP 3   Now fill in any values given in the problem. Use x for the unknown.

boxes	93	465
hours	2	x

STEP 4   If you can fill in three values with the fourth value missing, you can write and solve a proportion.

$93x = 465 \times 2$
$93x = 930$
$x = 930 \div 93$
$\boldsymbol{x = 10 \ hours}$

Now try the same steps above with this problem.

Admission to the flea market is $0.75 for children and $1.25 for adults. What would the admission costs be for the Sobala family of two adults and one child?

STEP 1   Find two things in the problem that are being compared.

cost for adults and cost for children

STEP 2   Make a chart and fill in the two things being compared.

Adults $		
Children $		

STEP 3   Now fill in any values given in the problem. Use x for the unknown.

Adults $	$1.25	
Children $	$0.75	

STEP 4   If you can fill in three values with the fourth value missing, you can write and solve a proportion.

You CANNOT fill in any more of the chart.

For the second problem, the proportion method will not work. You must find another way to solve the problem.

As you get more and more experienced with word problems, you will be able to see more quickly when to use a proportion.

··

Use the steps above to decide which of the problems can be solved using a proportion. Fill in as much of the chart with the labels and numbers as you can. Use x to stand for the unknown. Then check YES or NO. Do not solve the problems.

TIP
When you are given three numbers in a problem, and two of them have the same label (*dollars, boxes, feet, hours, dogs, shoes,* etc.), try writing a proportion. While it may not always work, it is worth a try.

WN 1. At Berry Book Sales there is one part-time employee for every 6 full-time employees. If there are 13 part-time employees in all, how many full-time employees are there?

label → | *part-time* | | |
label → | *full-time* | | |

Can this problem be solved using a proportion? YES____ NO____

D 2. Kathryn pays $15 to have 40 clean diapers delivered to her house. At this rate, what would she pay for 100 diapers?

label → | | | |
label → | | | |

Can this problem be solved using a proportion? YES____ NO____

F 3. Wendy's fruit salad recipe calls for 2 cups of yogurt for every 5 pounds of fruit. Wendy has only $1\frac{1}{2}$ cups of yogurt. How much fruit should she use?

label → | | | |
label → | | | |

Can this problem be solved using a proportion? YES____ NO____

P 4. During its busy season, the Oak Mall serves about 1,600 people each week. During the slow season, it serves about 70% of this number. How many people are served weekly in the slow season?

label →
label →

Can this problem be solved using a proportion? YES___ NO___

GE 5. Reed's father built a tree house that was 5 feet high, 5 feet wide, and 6 feet long. How many cubic feet was Reed's tree house?

label →
label →

Can this problem be solved using a proportion? YES___ NO___

D 6. Mark took his family of two adults and two children to the baseball game. If seats were $9.50 each, how much did Mark pay?

label →
label →

Can this problem be solved using a proportion? YES___ NO___

Go back and find answers to the problems on these two pages that *can* be solved by using a proportion.

Drawing a Picture

An electrician needed to cut several short pieces of wire from a spool that held 130 inches of wire. He wanted each piece to measure 5 inches long. How many pieces can he get from this spool?

Do you multiply or divide to find the answer?
How did you decide?

Sometimes you may know right away which operation to use to solve a word problem. Other times you may write a number sentence or equation that helps you decide. Here is another strategy that can help if you get stuck: You can draw a picture.

Here's a drawing of what is going on in the problem above.

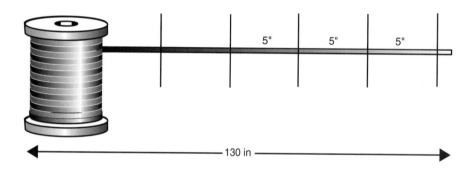

Did You Know . . . ?

- You don't have to be a very good artist to draw a simple sketch of a problem.
- You are the only one who has to be able to understand your drawing—it has to make sense only to *you*.
- A drawing is simply a tool that helps you visualize what is happening in a problem.

With the drawing above, you might find it easier to see that you should *divide* 130 by 5 to get your answer.

$130 \div 5 = $ **26 pieces**

Let's look at another problem.

The General K Company decided to make its new cereal box $\frac{1}{4}$ larger than its old box. If the old box held 32 ounces, how many ounces will the new box hold?

Draw a picture of what is happening in this problem. What operations will you need to use to solve it?

STEP 1 Start by drawing something.

STEP 2 Put in any numbers you know as labels.

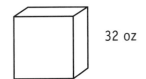

32 oz

STEP 3 Add to the drawing any more information given.

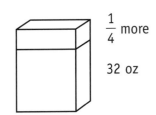

$\frac{1}{4}$ more

32 oz

STEP 4 Decide what operation to use.

Since $\frac{1}{4}$ is being added, I'll find $\frac{1}{4}$ of 32 and add it to 32.

STEP 5 Solve.

$\frac{1}{4}$ of 32 = 8
32 + 8 = 40

*The new box is **40 oz**.*

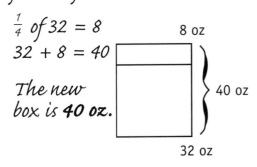

8 oz

40 oz

32 oz

Each of the following problems has a picture with it. Use the picture to solve the problem.

EXAMPLE   By 7:00 P.M. Tom's rain barrel had collected 2 inches of rain. By 9:00 P.M. the barrel contained 3 inches. How many inches had been collected between 7:00 and 9:00 that night?

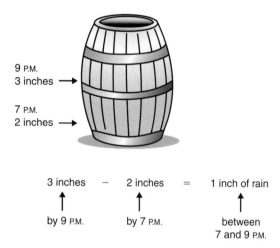

9 P.M.
3 inches →

7 P.M.
2 inches →

3 inches − 2 inches = 1 inch of rain
    ↑   ↑   ↑
 by 9 P.M.   by 7 P.M.   between
                           7 and 9 P.M.

WN 1. A political organization needs 960 signatures on its petition. If each petition holds 24 signatures, how many petitions need to be filled?

24 signatures ⌐

960 signatures

F 2. A worker spent the day driving steel posts into the ground. Each post is set into a hole that is $2\frac{1}{2}$ feet deep. If the post measures 8 feet tall, what is the aboveground height of the post?

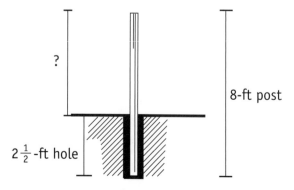

?

8-ft post

$2\frac{1}{2}$-ft hole

WN 3. The soup-kitchen volunteers need to use up 99 sticks of margarine before they spoil. If the sauce recipe they are using requires 3 sticks per pot, how many pots can they make?

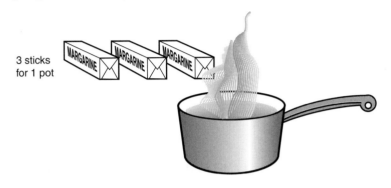

3 sticks for 1 pot

F 4. Of the 27 miles between home and work, Nigel has traveled halfway. How many more miles does he have to go?

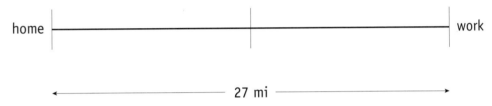

home work

27 mi

F 5. A seamstress shortened a skirt from $34\frac{1}{2}$ inches to $31\frac{1}{4}$ inches. How many inches did she take off the skirt?

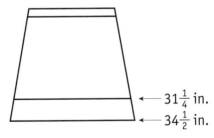

$31\frac{1}{4}$ in.
$34\frac{1}{2}$ in.

GE 6. A warehouse employee makes a stack of packing cartons that is 10 cartons high, 5 cartons wide, and 6 cartons deep. If each carton holds 8 cans, how many cans are in the stack?

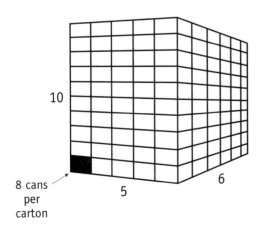

10

8 cans per carton

5

6

Solve the following word problems. Make a drawing to "picture" the problem.

WN 7. In one day Ricky's mother drove 6 miles north to school, then the same distance south to get home again. Next, she drove east to the grocery store 3 miles away and southwest to the nursing home, which was another 12 miles. Finally, she drove 9 miles home from the nursing home. How many miles did Ricky's mother drive in all?

F 8. Mrs. Antony knows that she needs $\frac{1}{2}$ can of soup for every pound of beef that she wants to stew. If she has 12 cans of soup in her cupboard, how many pounds of beef can she stew?

P 9. Lee knew that she had to plow a certain amount each day to get her fields prepared for planting. The first day her crew plowed 16 acres, and the next day the crew plowed 25% of what they had done on the first day. How many acres were plowed altogether?

GE 10. To buy the correct amount of paint for his room, Fred has to figure out the area of wall space. Each wall is square and measures 10 feet per side. What is the total are of all 4 walls?

GE 11. Estella makes hair scarves for extra money. She wants to put sequined trim along the edges of a triangular scarf that is 2 feet on each of 2 sides and 2 feet 4 inches on the third side. How much trim does she need?

GE 12. A rectangular field is twice as long as it is wide. If the field is 85 yards wide, how long is the field?

Mixed Review

Solve the following word problems. Use equations, proportions, pictures, and charts to help yourself.

WN **1.** A maintenance worker uses 6 gallons of white paint and 4 cans of varnish for every apartment he remodels. He has 2 apartments left to remodel. If he has only 2 gallons of white paint on hand, how many more will he need to buy to finish both apartments?

    a. 18   **d.** 4

    b. 10   **e.** not enough information given

    c. 8

WN **2.** A pharmacy worker had 250 milliliters of an antibiotic to mix with 400 milliliters of a flavored syrup. This mixture was then to be divided into small jars containing 50 milliliters. Which expression shows the number of jars the pharmacy worker could fill?

    a. $(250 \times 400) \times 50$ **d.** $400 - \dfrac{250}{50}$

    b. $\dfrac{250 \times 400}{50}$ **e.** $\dfrac{400 + 250}{50}$

    c. $\dfrac{400 - 250}{50}$

WN **3.** For every 2 ounces of infant formula concentrate, Veralee needs 6 ounces of sterilized water to feed her sick baby. If the can of concentrate contains 16 ounces, how many ounces of water will Veralee need?

    a. 8   **d.** 96

    b. 32   **e.** 108

    c. 48

WN **4.** According to the graph at the right, what was the average temperature for the week in Edgarton?

    a. 60°   **d.** 75°

    b. 65°   **e.** 525°

    c. 70°

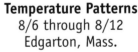

Temperature Patterns
8/6 through 8/12
Edgarton, Mass.

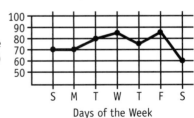

Temperature (in degrees)

Days of the Week

F 5. The map shows the route that Mrs. Nguyen takes when it is her turn for the car pool. If the total distance she travels from her home to the school is $7\frac{1}{2}$ miles, what is the distance from Theo's house to the school?

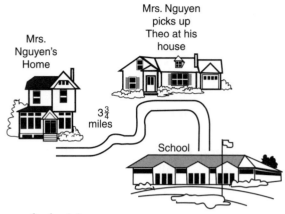

Mrs. Nguyen's Home

Mrs. Nguyen picks up Theo at his house

$3\frac{3}{4}$ miles

School

a. $8\frac{1}{4}$ d. $3\frac{3}{4}$

b. $6\frac{1}{2}$ e. 2

c. $4\frac{1}{2}$

P 6. In 1998 the average American watched 53.5 hours of television per week. Approximately 30% of this time is taken up by advertising. About how many hours per week is the average American watching commercials on TV?

a. 16.05   d. 19.5

b. 16.5   e. 160.5

c. 17.85

 Problems 7–8 refer to the following information.

Rachel and Martha are purchasing a home together. Because Martha earns more money than Rachel, they have agreed that Martha will pay 60% of the mortgage payment and Rachel will pay 40%. They are equally dividing $12,000 as a down payment, and their monthly mortgage payments will be $750. In addition, they estimate that their utility bills will total $180 each month. Since Rachel works at home, she will pay for all utilities.

WN 7. Martha had $13,350 in her savings account. After she withdraws her share of the down payment, how much will she have remaining in the account?

a. $1,350   d. $12,000

b. $6,000   e. not enough information given

c. $7,350

P 8. How much money will Rachel pay each month, including mortgage and utilities?

a. $72   d. $480

b. $300   e. $4,980

c. $372

SOLVING THE PROBLEM

Keeping Organized

Jan works in a day-care center 20 hours a week and earns $7 per hour there. At night she works as a waitress in a restaurant, where she earns twice that amount each week. How much money does Jan earn altogether in a week?

a. $20
b. $140
c. $280

d. $420
e. not enough information given

What do you need to do to get the right answer to this problem?

Use the space below to solve the problem. When you are finished, close this book and wait five minutes before you open it again.

Have you waited 5 minutes? If so, look at your work in the space and do the following:

- Circle the number of hours Jan works each week at the day-care center.

- Underline the amount of money she earns weekly at the day-care center.

- Put a box around the amount of money Jan earns weekly as a waitress.

- Put a star next to your answer to the problem.

Was it easy for you to find all the numbers? Or did you have to search around the page and rethink the problem in your head?

Believe it or not, many mistakes are made because people do not work in an organized way. Look at the following solution to the problem above, and go through the steps listed. Notice how easy it is to follow the work.

Day Care	Waitress	
$7 per hour	$140	$280
× 20 hours	× 2	+ $140
$140 each week	$280 each week	$420 total

If you are getting the right answer most of the time, you have probably found a process that works well for you. However, if you think you should be getting the right answer more often, and you are unsure about what you are doing wrong, keep in mind the following.

TO STAY ON THE PATH TO THE RIGHT ANSWER . . .

1. Be sure to line up your numbers neatly.

   THIS $280   IS NOT THIS $280
          + 140   + $120
          ─────   ───────
           $420 $2,920

2. Keep each step separate and number the steps if it helps.

   THIS   **STEP 1** $7.00   IS BETTER THAN THIS
                       × 20 $7.00
                      ───────   × 20
                      $140.00   ───────
                                                  140
          STEP 2 $140   140
                     × 2   2
                     ─────   ─────
                      $280   280

3. Use labels whenever possible.

   THIS WILL $280   (waitress)
   NEVER   +140   (day care)
   CONFUSE YOU   ────
                   $420   (total)

   AS THIS   280   (hours?? days?? day care??)
   WILL   +140   (dollars?? hours?? waitressing??)
                   ────
                   420   (hours?? day?? waitressing??)

> **TIP**
> Working in an organized way won't always ensure the right answer, but sloppy and careless work will almost always lead to mistakes.

4. Make sure that the numbers you have written are the same as the numbers in the problem.

   THIS $7 per hour, 20 hours

   CAN BE COPIED INCORRECTLY AS $20 per hour, 7 hours

First solve each problem below using any hints from above that you think might help you. Then read the student's solution that follows each problem. Find the mistake that led to the incorrect answer.

EXAMPLE   Minni started a business as a carpenter. On weekdays she works 9 hours a day. On Saturdays she works 4 hours. How many hours does she work per week?

a. **Correct Solution**

$$\begin{array}{r} 9 \text{ hours} \\ \times\, 5 \text{ days} \\ \hline 45 \text{ hours} \end{array} \qquad \begin{array}{r} 45 \text{ hours} \\ +\, 4 \text{ hours} \\ \hline 49 \text{ hours} \end{array}$$

b. **Incorrect Solution** What is the mistake in this solution?

$$\begin{array}{r} 9 \text{ hours} \\ +\, 4 \text{ hours} \\ \hline 13 \text{ hours} \end{array}$$

The student forgot to multiply the 9 hours by the 5 weekdays.

WN 1. The chart on the right shows the number of hits the Billtown baseball team got in last week's games. What is the average number of hits for the best 3 hitters?

Player	Hits
Fernando Benson	12
Jim Ryan	1
David Burns	4
Tyson Jones	11
Lewis Caviness	8
Doug Mills	10
John McCarthy	9
Mic Thomson	9

a. Your solution:

b. What is the mistake in this solution?

$12 + 1 + 4 + 11 + 8 + 10 + 9 + 9 = 64$

$64 \div 8 = 8$ hits

The mistake is

F 2. Michael, the neighborhood handyman, mowed 4 lawns, fixed 2 clogged drains, and worked for 8 hours on a remodeling project this week. If each lawn took him $1\frac{1}{2}$ hours to mow and each drain took 1 hour to fix, how many hours in all did he work?

    a. Your solution:

    b. What is the mistake in this solution?

        STEP 1 $4 \times 1\frac{1}{2} = 6$ hours mowing

        STEP 2 $2 \times 1 = 2$ hours fixing drains

        STEP 3 $6 + 2 = 8$ hours

        _____

        _____

M 3. Louisa is making chili for her teenagers and has decided to triple her recipe. The original recipe calls for 2 pounds 8 ounces of canned kidney beans and 2 pounds of hamburger. How much kidney beans and hamburger should she buy to triple the recipe?

    a. Your solution:

    b. What is the mistake in this solution?

        STEP 1 $2.8 \times 3 = 8.4$ pounds of kidney beans

        STEP 2 $2 \times 3 = 6$ pounds of hamburger

        _____

        _____

Estimating the Answer

Tara waitressed Saturday and earned $15.75 in tips. The previous Saturday she earned 3 times that much. How much money did Tara earn on the previous Saturday?

a. $5.25   **b.** $12.75   **c.** $31.50   **d.** $47.25   **e.** $57.20

Without doing any computation, decide which of the answer choices above *might* be the right answer.
How did you decide?

When you **estimate** an answer to a problem, you get a rough idea of the correct answer by rounding off the numbers in the problem to make them easier to work with. Here are some steps to follow.

$15.75 is about $16

STEP 1   Try an estimated solution.

STEP 2   Choose the answer choice closest to your estimate.

STEP 3   Check your estimate with the original numbers.

$16 × 3 = $48 (estimated answer)
*Choice **d.** $47.25 is closest to $48.*

*$15.75 × 3 = **$47.25** (real answer)*

On multiple-choice tests, an estimate helps you choose the correct answer quickly.

You can indicate "is approximately equal to" with the sign ≈. In the problem above, $47.25 ≈ $48.

For each problem below, first estimate an answer. Then use your estimate to choose from the list of answers. Finally, find an exact answer and compare it to your estimate.

EXAMPLE   An assembly-line worker can complete 864 circuit boards in 24 hours. At this rate, how many boards can she complete in 1 hour?

a. 28   b. 30   c. 36   d. 360   e. 400

Estimate: 864 → 900 circuit boards   900 ÷ 20 = **45**
                24 → 20 hours

Answer choice: c. 36   36 is closest to 45.

Exact solution: 864 ÷ 24 = **36**   This verifies the estimate.

WN **1.** Harold weighs 68 pounds more than his mother, who weighs 41 pounds more than Harold's sister. If Harold weighs 211 pounds, what does his sister weigh?

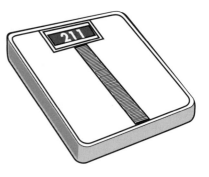

   a. 170   **b.** 143   **c.** 114   **d.** 102   **e.** 62

   Estimate: 211 → _____   **Answer choice:**   **Exact**
   68 → _____   **solution:**
   41 → _____

D **2.** Hamburger meat is on sale at Harold's Market for $2.19 per pound. How much would a package weighing 3.25 pounds cost?

   a. $4.30   **b.** $7.12   **c.** $10.42   **d.** $11.18   **e.** not enough information given

   Estimate: **Answer choice:**   **Exact solution:**

D **3.** Regular gas at Fred's Fill-Up Station costs $1.14 per gallon. How much did Doreen pay for 10.8 gallons?

   a. $6.64   **b.** $8.35   **c.** $9.01   **d.** $12.31   **e.** $18.30

   Estimate: **Answer choice:**   **Exact solution:**

P **4.** A total of 403 students participated in the sports program last year. What percent of the student body participated if there are 2,015 students at the school?

   a. 5%   **b.** 15%   **c.** 20%   **d.** 81%   **e.** 160%

   Estimate: **Answer choice:**   **Exact solution:**

GE **5.** Joe's garden is 8 yards 8 inches wide and 32 yards long. How much fencing should he buy to completely surround his garden?

   a. 8 yd 16 in.   **b.** 256 yd   **c.** 40 yd 8 in.   **d.** 64 yd

   Estimate: **Answer choice:**   **Exact solution:**

Using Friendly Numbers to Estimate

Loyal Landscaping Company delivered a total of 95 cubic yards of bark mulch to 16 different sites. About how many yards went to each site?

a. 4   **b.** 6   **c.** 8   **d.** 16   **e.** 21

One way to solve this problem is to divide 16 into 95. However, there is an easier way—a way that will save you time when you are taking a test. You can estimate, using "friendly" numbers.

Friendly numbers are numbers that are easy to work with. They are numbers that can be added, subtracted, multiplied, or divided in your head. For example, in the problem above, what number is close to 95 that would be easily divisible by 16?

$95 \div 16$ is very close to $96 \div 16$

Since $96 \div 16$ is 6, you know that $95 \div 16$ is pretty close to 6 as well. You can pick answer choice **b. 6** without doing any pencil and paper computation.

Let's look at another example.

Michael paid $5.40 for 11 pads of paper. How much did each pad cost?

a. $1.10   **b.** $0.90   **c.** $0.85   **d.** $0.49   **e.** $0.33

To solve this problem without any messy computation, follow these steps.

STEP 1 Find an amount close to $5.40 that will divide easily by 11.

The amount $5.50 is pretty close to $5.40, and it can easily be divided by 11.

STEP 2 Divide using the friendly numbers.

$5.50 ÷ 11 = $0.50.

STEP 3 Look at the answer choices and find one that is close to your estimate.

*Answer choice **d. $0.49** is close to $0.50.*

For each problem below, first find a pair of friendly numbers to use for an estimate. You may change one or both numbers in the problem to make them easy to work with. Then choose the answer closest to your estimate.

EXAMPLE   A deli worker puts 34 pints of cole slaw into containers holding 1.8 pints. How many full containers did she have in all?

   a. 2   **b.** 8   **c.** 18   **d.** 30   **e.** 42

   Friendly numbers: 34 becomes <u>34</u>; 1.8 becomes <u>1.7</u>

   34 ÷ 1.7 = 20   **c. 18** is closest

1. Wally cut 2.9 meters of tubing into 1.4-meter lengths. How many pieces did he have in all?

   a. 2   **b.** 3   **c.** 4   **d.** 8   **e.** 10

   Friendly numbers: 2.9 becomes_____; 1.4 becomes_____

   Closest answer choice:_____

2. What is 208 pounds divided by 7.2?

   a. 24.1   **b.** 28.9   **c.** 34.8   **d.** 40.2   **e.** 46

   Friendly numbers: 208 becomes_____; 7.2 becomes_____

   Closest answer choice:_____

3. One inch is equal to 2.54 centimeters. How many whole inches are there in 100 centimeters?

   a. 15   **b.** 20   **c.** 24   **d.** 32   **e.** 39

   Friendly numbers: 100 becomes_____; 2.54 becomes_____

   Closest answer choice:_____

Estimating with Fractions

The cinder blocks that Gayle was moving weighed $3\frac{1}{4}$ pounds each. If the pallet he used could hold no more than $48\frac{3}{4}$ pounds, what was the maximum number of blocks Gayle could place on it?

a. 158   **b.** 45   **c.** 15   **d.** 13   **e.** 3

Imagine this problem without fractions. How would you solve it?

Fractions can often be difficult to work with. Sometimes just looking at fractions can confuse you! Using estimation with fractions can often help you see how to solve a problem.

TIP
Use rounded numbers instead of fractions to help you see more clearly how to solve the problem.

Here are some steps to help you estimate with fractions.

STEP 1   Round all numbers with fractions.

- If the fraction is greater than or equal to $\frac{1}{2}$, drop the fraction and add 1. (round up)

    $48\frac{3}{4} \approx 49$ ($\frac{3}{4}$ is greater than $\frac{1}{2}$)

- If the fraction is less than $\frac{1}{2}$, just drop it. (round down)

    $3\frac{1}{4} \approx 3$ ($\frac{1}{4}$ is less than $\frac{1}{2}$)

STEP 2   Solve the problem using the estimated numbers.

$49 \div 3 = 16\frac{1}{3}$
(Or round down to 48 because 3 divides evenly into 48)
$48 \div 3 = \textbf{16}$

STEP 3   Find the answer choice closest to your estimate.

*Choice **c. 15** is closest.*

STEP 4   You can check your estimate with the original numbers.

$48\frac{3}{4} \div 3\frac{1}{4} = \textbf{15}$
You estimated the correct answer.

For each problem below, first estimate an answer using the steps outlined above. Use your estimate to choose from the list of answers. Then find the exact answer and compare it to you estimate.

F 1. One week Flora worked $24\frac{1}{2}$ hours at a rate of $9.70 per hour. What were Flora's earnings that week?

    a. $2,376.50   **b.** $237.65   **c.** $200.65   **d.** $24.25   **e.** not enough information given

    Estimate: Answer choice: Exact solution:

F 2. What is the difference in length (in inches) between the two rods shown at the right?

24 5/8 "

10 1/4 "

    a. 35   **b.** $34\frac{1}{8}$ **c.** $14\frac{3}{8}$ **d.** 12   **e.** not enough information given

    Estimate: Answer choice: Exact solution:

F 3. Josie needs to figure out how many teaspoons of sugar are in the muffins she just ate. If she ate $3\frac{1}{2}$ muffins, and each muffin contains $2\frac{1}{4}$ teaspoons of sugar, how many teaspoons of sugar did she eat in all?

    a. 2   **b.** 4   **c.** $7\frac{7}{8}$ **d.** 10   **e.** $12\frac{1}{2}$

    Estimate: Answer choice: Exact solution:

F 4. A dressmaker had fabric remnants measuring $2\frac{1}{5}$ yards, $3\frac{1}{4}$ yards, $3\frac{3}{10}$ yards, and $4\frac{7}{10}$ yards. How much fabric did she have in all?

    a. $13\frac{9}{20}$ **b.** $10\frac{3}{10}$ **c.** $8\frac{1}{4}$ **d.** $6\frac{1}{5}$ **e.** $5\frac{1}{2}$

    Estimate: Answer choice: Exact solution:

When an Estimate Is the Answer

The Lanier family spends $122 per week on groceries, while the Lake family spends $39. About how many times more do the Laniers' groceries cost than the Lakes'?

a. $\frac{1}{2}$ **b.** 2 **c.** 3 **d.** 4 **e.** 5

How do you know when to estimate the answer to a problem?
Do you need to compute an exact answer for the problem above?

Word problems like the one above do not require you to find an exact answer. The clue that you need only an *estimated* answer for the problem above is the word *about*. In other problems you may see the word *approximately*. These clue words tell you to round the numbers in the problem to get an *approximate* answer.

Round the numbers above.

$122 \approx 120$ $39 \approx 40$

By looking quickly at these rounded numbers, you can see that 120 is 3 *times* 40.

$120 \div 40 = 3$

The answer to this problem is **c. 3** *even though* 3×39 is not *exactly* 122. Remember that the question asked "*about* how many times?"

> **TIP**
> Look for clue words such as *about* and *approximately*. When you see these words in a word problem, think about estimating the answers.

Solve the following word problems by estimating. Remember to watch out for clue words.

WN 1. The first week on her new job as a salesperson, Sandy made 4 sales. By the last week of the month she made 23 sales. About how many times did her sales increase?

    a. 4   **b.** 6   **c.** 19   **d.** 23   **e.** not enough information given

M 2. It takes Earl about 7 minutes to run around the school yard. About how many times will he need to run around the yard to complete his 1 hour of physical education?

    a. 1   **b.** 4   **c.** 8   **d.** 60   **e.** not enough information given

Problems 3–5 refer to the graph shown below.

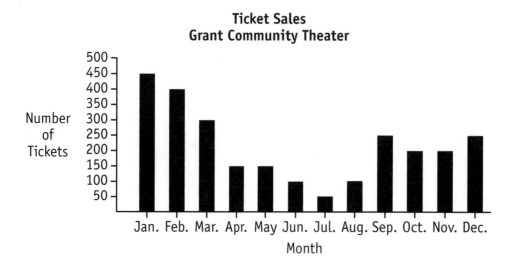

Ticket Sales
Grant Community Theater

WN 3. About how many times as many tickets were sold in January as in July?

    a. 9   **b.** 10   **c.** 50   **d.** 450   **e.** not enough information given

D 4. Tickets cost $3.75 each. About how much money did the theater take in during the first 3 months of the year?

    a. $300   **b.** $400   **c.** $450   **d.** $1,150   **e.** $4,600

P 5. Approximately what percent of the April–May–June sales took place in June?

    a. 1%   **b.** 5%   **c.** 10%   **d.** 25%   **e.** 50%

Writing Answers in Set-Up Format

To be fair, Mr. Méndez collects all customer tips and takes $10 out for each of his 2 busboys. He then divides the remainder equally among his 3 waitresses. If he collected $134 in tips, which expression shows the amount received by each waitress?

a. $\dfrac{\$134 - \$2}{3}$

b. $\dfrac{\$134 - 10}{3}$

c. $(\$134 \div 3) - 10$

d. $(\$134 \div 3) - \$20$

e. $\dfrac{\$134 - \$20}{3}$

Do you remember working with set-up problems? How do you go about finding the right answer?

TIP
If you are having trouble recognizing answers in set-up problems, decide whether your set-up can be written another way. Remember, there is often more than one way to set up a problem.

In the chapter called Understanding the Question, you learned about set-up questions and how they are different from other word problems. In this lesson you will again look at set-up problems and see different ways to find a solution.

To solve the problem above, follow the steps you learned on page 47.

STEP 1 Make a statement that tells what needs to be done.

I need to subtract the busboys' pay from the total tips, then divide by the number of waitresses.

STEP 2 Fit in the numbers, just as you would write a plan for a number sentence.

I need to take two $10 amounts from $134, then divide by 3.

STEP 3 Write an expression or a number sentence.

$$\dfrac{\$134 - (2 \times \$10)}{3}$$

STEP 4 Find the answer choice that matches your set-up.

At first you may think that no answer matches the expression in Step 3. The expression $\dfrac{\$134 - (2 \times \$10)}{3}$ is not listed as one of the choices. However, look again.

$\dfrac{\$134 - (2 \times \$10)}{3}$ can also be written as $\dfrac{\$134 - \$20}{3}$ (answer choice **e.**)

Did You Remember . . . ?
- Parentheses ( ) mean "do me first."
- A fraction bar (—) or slash bar (/) means "divided by" (÷).
- A number written outside parentheses means multiplication. $35(230) = 35 \times 230$
- Writing set-up expressions is a lot like writing equations.

For each problem below, write *two* different set-up expressions. To indicate multiplication, you may use either × or ( ); to indicate division, use either ÷ or /.

> **TIP**
> Practice writing set-up expressions as you solve regular word problems. This will help you work with the set-up format, **and** it will help you solve the problems.

EXAMPLE   Agatha earns $5.75 per hour cleaning offices. She worked 8 hours on Monday, 8 hours on Thursday, and 10 hours on Saturday. Find Agatha's pay for the 3 days.

    a. *$5.75(8 + 8 + 10)*

    b. *8($5.75) + 8($5.75) + 10($5.75)*

WN **1.** Find the distance Denny traveled if he drove for 3 hours at 55 miles per hour and another 3 hours at 60 mph.

D **2.** At a garage sale, Val spent a total of $32.80. One of her purchases was a pair of candlesticks that cost $3.00 per candlestick. She spent the rest of the money on a blanket. Find the cost of the blanket.

F **3.** At Cecil's Cleaning Service, every employee is expected to work 36 hours per week. Over the weekend, Byron worked $\frac{1}{4}$ of his expected hours. Find how many more hours Byron needs to work this week.

GE **4.** Find the distance around the backyard shown at the right.

20 ft

40 ft

Choosing the Correct Expression

David pays a total of 28% of his yearly salary in taxes and other deductions. Which of the following expressions shows David's yearly take-home pay if his gross salary is n?

a. $0.28n$ **b.** $n + 0.28n$ **c.** $n - 28n$ **d.** $n - \frac{n}{28}$ **e.** $n - 0.28n$

What steps can you use to find the correct expression?

Some set-up questions use a letter, or a **variable,** to represent a number we do not know. Using a variable means that no matter what number you substitute for the letter, the given expression will provide the correct value.

For example, in the problem above, suppose David's gross salary is $37,000. What would his take-home salary be after taxes and other deductions? To find his take-home salary, you probably know that you have to subtract taxes and deductions from his gross salary. You know that 28% of his gross salary should be subtracted.

gross salary   total deductions
$\downarrow$ $\downarrow$
$37,000 $-$ 0.28($37,000)   **Remember:** Parentheses mean multiplication.

Now suppose you don't actually know David's gross salary. You decide to just call it n—a letter than can stand for any number. Which expression above shows David's take-home pay if n is his gross salary?

You're right if you chose **e. $n - 0.28n$.**

Now use the steps to find the correct expression for the problem below.

Mrs. Fong receives time-and-a-half pay for each hour she works overtime at her new job. This means that she receives one and a half times her regular hourly wage for each hour of overtime she works. If Mrs. Fong's regular pay is n dollars per hour, which expression shows the amount of money she will earn in 6 hours of overtime work?

a. $6n$ **b.** $0.5(6n)$ **c.** $6(1.5n)$ **d.** $6n + 1.5n$ **e.** $\frac{6}{1.5n}$

STEP 1	Find Mrs. Fong's pay per hour of overtime work.	*The problem states that she earns $1\frac{1}{2}$, or 1.5, times her regular salary of n. 1.5n would be her overtime wage per hour.*
STEP 2	Now find Mrs. Fong's overtime pay if she works six hours.	*$6 \times 1.5n$ is her overtime pay.*
STEP 3	Find an answer choice that gives the same value as $6 \times 1.5n$.	*c. 6(1.5n) is the same as $6 \times 1.5n$.*

Choose the correct expression for each problem below. Use the steps outlined above to help you.

1. A customer bought a shovel at the hardware store. He also bought a rake that cost twice as much as the shovel. If x equals the amount he paid for the shovel, which expression shows the amount he spent at the hardware store?

   a. $\frac{x}{2}$　　**b.** $3x$　　　　**c.** $2x$　　　　**d.** $x + \frac{1}{2}x$　　**e.** $2x + \frac{1}{2}x$

2. John puts x dollars into a savings account each week. At the end of the month, he withdraws \$475 to pay his rent. How much money is in his account after four weeks?

   a. $x - 475$　　**b.** $4x - (4 \times 475)$　　**c.** $\frac{(x - 475)}{4}$　　**d.** $4x - 475$　　**e.** $475x$

3. A student received scores of 80, 78, and 92 on her math tests. She has not received her score on her last test. If y equals the score on the last test, which expression shows her average score for all the tests?

   a. $\frac{(80 + 78 + 92)}{3}$　　　　**c.** $\frac{(80 + 78 + 92 + y)}{4}$　　　　**e.** $4(80 + 78 + 92 + y)$

   b. $\frac{(80 + 78 + 92)}{y}$　　　　**d.** $\frac{(80 + 78 + 92 + y)}{3}$

4. If g equals the cost of a dozen donuts, how much does one donut cost?

   a. $11g$　　**b.** $\frac{g}{10}$　　**c.** $12g$　　**d.** $\frac{g}{144}$　　**e.** $\frac{g}{12}$

Comparing and Ordering Numbers

A survey showed that $\frac{2}{5}$ of the workers in Company A supported the union. In Company B, 55% supported it, and 75% of Company C supported it. Which of the following sequences shows the order of companies from *greatest* support to *least* support?

a. A, B, C   **d.** B, A, C

b. C, B, A   **e.** C, A, B

c. B, C, A

What do you need to do to solve this problem?
How can you compare fractions and percents?

Many word problems and real-life situations require you to compare numbers. This is easier when you are comparing 35 truckloads with 1,000 truckloads or 2 hours with 2 hours. But what about when the **labels** are not the same?

Here are some steps you can take when you are asked to compare and order numbers.

STEP 1	Choose *one* label for all the numbers you must compare. Try to choose the label that will make the least work for you.	*In the problem above, you could change all numbers into fractions or percents. Since two of the numbers are already percents, choose percent.*
STEP 2	Carefully change all numbers so that they have the same label.	*Company A: $\frac{2}{5}$ = 40%* *Company B: 55%* *Company C: 75%*
STEP 3	Now put the numbers in the correct order.	*This problem asks for largest to smallest: 75%, 55%, 40%*
STEP 4	Match the letters with the numbers.	*75% — C* *55% — B* *40% — A* *The correct answer is **b. C, B, A***

Now try another type of comparison problem.

Based on the salaries listed below, Mathias is deciding which job to take.

A. $24,000 per year, with a 10% raise for the second year
B. $12 per hour, 40 hours per week, with no raise
C. $10 per hour, 40 hours per week, with a $1,000 bonus
D. $26,000 per year, with a 5% raise for the second year

Which job will give Mathias the most money in his second year of employment?

a. Job A
b. Job B
c. Job C
d. Job D
e. not enough information given

> **TIP**
> When you do a comparing or an ordering problem, **always look at the labels first.** Make the necessary conversions; then compare.

STEP 1 Decide how to convert numbers so that they can be compared.

You need to convert numbers to dollars for the second year.

STEP 2 Carefully perform all computations.

JOB A
10% of $24,000 = $2,400
$24,000 + $2,400 = $26,400

JOB B
$12 × 40 hours = $480 per week
$480 × 52 weeks = $24,960

JOB C
$10 × 40 hours = $400 per week
$400 × 52 weeks = $20,800
$20,800 + $1,000 bonus = $21,800

JOB D
5% of $26,000 = $1,300
$26,000 + $1,300 = $27,300

STEP 3 Choose the correct answer.

d. Job D
because $27,300 is larger than all of the other second-year salaries

First decide what label you will use to convert the values in each problem below. Then do the necessary conversions and solve the problem.

P 1. Tony will buy whatever automobile costs him the least. Based on the information below, which car will Tony buy?

    A. A working 1994 Chevrolet for $9,600 cash.
    B. A new Hyundai for $11,000, with a $500 rebate.
    C. A 1998 Ford for $2,000 down and $200 per month for 3 years.
    D. A 1995 Ford for $6,000, with an added $1,500 for repairs.
    E. A 1999 Toyota for $12,000, less a 10% dealer discount.

LABEL TO CONVERT ALL VALUES TO: _____

ANSWER: _____

    a. A   **b.** B   **c.** C   **d.** D   **e.** E

P 2. A survey of five southwestern towns showed the following results regarding the question of whether or not a nuclear power plant should be built in the area.

> Town A: 62% of residents in favor
> Town B: $\frac{2}{3}$ of residents in favor
> Town C: 46% of residents opposed
> Town D: 48% of residents opposed
> Town E: $\frac{1}{3}$ of residents in favor

Which of the following lists the towns in order of *most in favor* to *least in favor?*

LABEL TO CONVERT ALL VALUES TO: _____

ANSWER: _____

    a. B, A, C, D, E
    b. A, C, D, B, E
    c. E, C, D, A, B

    d. C, E, A, D, B
    e. B, D, C, E, A

M 3. A carpenter has the following lengths of boards. To incur the least amount of waste, he wants to use the boards in order from *longest* to *shortest.* Which of the following lists the correct order?

Board A: $2\frac{1}{2}$ feet

Board B: 1 yard

Board C: 40 inches

Board D: $1\frac{1}{2}$ yards

Board E: $3\frac{1}{2}$ feet

LABEL TO CONVERT ALL VALUES TO: _____

ANSWER: _____

   a. E, D, C, A, B
   b. D, E, C, B, A
   c. C, D, B, A, D
   d. B, A, D, C, E
   e. A, B, D, C, E

M 4. A salad preparer had the following measures of cooking oil in the kitchen pantry. Which of the following lists their order of amount from *least* to *greatest?*

1 pint = 2 cups		1 cup = 8 ounces

   A. $1\frac{1}{2}$ cups of Wesson
   B. $\frac{1}{2}$ of an 8-ounce bottle of Puritan
   C. 1 pint of Ready-Pour
   D. 3 cups of a generic brand
   E. 6 ounces of Home's Cook

LABEL TO CONVERT ALL VALUES TO: _____

ANSWER: _____

   a. A, D, C, B, E
   b. D, C, B, E, A
   c. E, D, A, C, B
   d. B, E, A, C, D
   e. C, E, B, D, A

Making Good Use of Your Calculator

A survey shows that three-fourths of the households in the town are connected to town sewers. There are 1,280 houses north of the river and 2,840 houses south of the river. How many houses in all are connected to town sewers?

a. 3,090   **b.** 4,120   **c.** 1,260   **d.** 6,180   **e.** 2,080

How can you use a calculator to help solve this problem?

You may have figured out that you need to add the houses north and south of the river, then find three-fourths of that total.

$\frac{3}{4} \times (1{,}280 + 2{,}840) = $ houses on town sewer

One way to do the computation is to add the numbers in parentheses and multiply by $\frac{3}{4}$. But if you can use a calculator, there is a faster and probably more accurate way.

How can you work with fractions on a calculator? Simply by changing the fraction to a decimal. Perhaps you already know that $\frac{3}{4}$ is the same as 0.75. If you did not, there is a simple rule for changing a fraction to a decimal.

> To change a fraction to a decimal, divide the denominator into the numerator.

$$
\begin{array}{r}
0.75 \\
4\overline{)3.00} \\
\underline{2\ 8} \\
20 \\
\underline{20} \\
\end{array}
$$

Here's how you can solve the problem above using a calculator.

Press these keys:

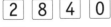 1 2 8 0

+

2 8 4 0

=

×

. 7 5

=

The calculator displays:

1280.
1280.
2840.
4120.
4120.
0.75
3090.

Use a calculator to solve the following problems. Remember that to change a percent to a decimal, move the decimal point two places to the left and drop the percent sign.

EXAMPLE Tawanna spends 40% of her work day caring for her neighbor's children. If she works a total of 50 hours per week, how many hours are spent doing childcare?

    a. 10   **b.** 15   **c.** 20   **d.** 30   **e.** 40

    40% = 0.40   0.40 × 50 = **20 hours**

1. Ricardo ran $1\frac{3}{4}$ miles one day, $2\frac{1}{2}$ miles the next, $4\frac{3}{4}$ the following day, and $1\frac{1}{4}$ the last day. How many miles in all did he run?

    a. 9.25
    b. $9\frac{1}{2}$
    c. $10\frac{1}{4}$
    d. 10.50
    e. $11\frac{3}{4}$

2. A newspaper reports that 17% of its readers receive the paper free of charge. If the newspaper has 22,500 readers, how many get the paper free?

    a. 22,483
    b. 13,235
    c. 10,280
    d. 6,450
    e. 3,825

3. How many times can a $3\frac{1}{2}$-minute commercial be shown in 49 minutes?

    a. 14
    b. 15
    c. 17
    d. 19
    e. 20

4. What is the area of this figure, in square feet?

    a. 5.25
    b. 64
    c. 107.5
    d. 245.125
    e. 250.5

13.24 ft / 18.5 ft

What to Do with Remainders

Problem 1

At a construction site, the crew dug up 13,500 cubic yards of sand and soil. If a truck can haul 200 cubic yards at a time, how many trips will the crew have to take to clear the site?

What operation do you use to solve this problem?
What is the answer?

You would need to divide to find the answer to this problem.

$$
\begin{array}{r}
67\,\text{R}100 \\
200\,\overline{)13,500} \\
\underline{12\ 00} \\
1\ 500 \\
\underline{1\ 400} \\
100
\end{array}
$$

> **TIP**
> Whenever you find a remainder in your computation, READ THE QUESTION AGAIN. The question will give you a clue whether to round the remainder up or down.

But what is the answer to the problem? 67? 67 remainder 100? $67\frac{1}{2}$?

Notice that the question above asks for *the number of trips* necessary to remove *all* the soil. After 67 trips, there are still 100 cubic yards left. The crew will need to take another trip.

The answer to the problem is **68 trips.** You needed to *round up*.

Let's look at another remainder problem.

Problem 2

A meat-packing company uses 40-pound crates to ship its prime cuts. If a packer has 4,300 pounds of prime beef, how many *full* crates can he ship?

a. 10   **b.** 100   **c.** 107   **d.** $107\frac{1}{2}$ **e.** 108

What should you do with the remainder in this problem?
How do you know?

STEP 1   Divide.

$$40\overline{)4,300}^{107\,R\,20}$$
$$\underline{40}$$
$$300$$
$$\underline{280}$$
$$20$$

STEP 2   Reread the question.

"How many full crates?"

STEP 3   Decide what to do with the remainder.

Since the 20 pounds left over cannot make a full crate, do not use the remainder. You round down.

STEP 4   Choose your answer.

The packer can ship 107 crates. **c. 107**

What to Do with Remainders

Round Up	Round Down
• When you **need to use** the remainder	• When you **don't need to use** the remainder

In problem 1, you needed to know how many trips—including partial loads.

In problem 2, you needed to know how many *full* crates only.

Solve the following word problems. Pay special attention to remainders.

EXAMPLE   A land development company owns 1,560 acres of land in Fairfield County. It plans to sell the land in 7-acre plots. How many of these 7-acre plots can be sold?

Answer with remainder: *222 R6*

Round   UP or (DOWN)   (circle one)

Why? *The question asks for the number of*

7-acre plots. The 6 acres are left over.

Answer: *222*

WN 1. The Carson Bus Lines has agreed to transport a group of schoolchildren to and from the zoo. The school is sending 6 classrooms of 30 students per class. If each bus can hold 40 students, how many buses will be needed?

Answer with remainder: _____

Round   UP or DOWN   (circle one)

Why? _____

_____

Answer: _____

WN 2. Mathilda needs 16 yards of fabric for her bridesmaids' dresses. The fabric she has chosen is being sold as remnants cut into 3-yard lengths. How many pieces will Mathilda need to buy?

Answer with remainder: _____

Round   UP or DOWN   (circle one)

Why? _____

_____

Answer: _____

F 3. Bert has $10\frac{1}{2}$ cups of oatmeal to use in a cookie recipe. If one batch of cookies requires $2\frac{1}{2}$ cups of oatmeal, how many complete batches can Bert make?

Answer with remainder: _____

Round   UP or DOWN   (circle one)

Why? _____

_____

Answer: _____

Mixed Review

Choose the correct answers for the following problems.

Problems 1–3 refer to the following graph.

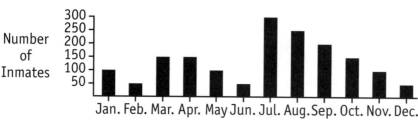

WN **1.** What was the average number of inmates at Franklin County during the first half of the year?

    a. 600   **b.** 300   **c.** 200   **d.** 100   **e.** 50

WN **2.** In February, how many Franklin County inmates were charged with drug offenses?

    a. 5   **b.** 10   **c.** 25   **d.** 50   **e.** not enough information given

P **3.** A total of 69% of the Franklin County inmates in July were repeat offenders. Which expression shows the number of repeat offenders in July?

    a. $0.69(100)$ **d.** $0.69 + 300$
    b. $300 - 0.69(300)$ **e.** $0.69(300)$
    c. $0.69 \div 300$

D **4.** Your car's gas tank can hold 14 gallons of gas. The gas gauge measures $\frac{1}{2}$ full. About what will it cost to fill up the tank if gas costs $1.17 per gallon?

    a. $.80   **d.** $32
    b. $8   **e.** not enough information given
    c. $16

F 5. Rafael has $6\frac{1}{2}$ more hours to work today. He worked 9 hours yesterday and 10 hours the day before. What more do you need to know to find the total hours Rafael worked in the 3 days?

    a. his hours per week
    b. his wages per hour
    c. the hours he has already worked today
    d. his average daily hours
    e. the time he got to work today

P 6. Of the 4,450 residents of Honeysuckle Hill, 40% say they will vote for Matthew Locke for mayor and 30% say they will vote for Gary Meyers. The rest of the residents are undecided. Which expression shows the number of undecided residents?

    a. $4,450 + 0.70(4,450)$ **d.** $4,450 + 0.4(4,450)$
    b. $4,450 - 0.70(4,450)$ **e.** $4,450 - 0.4(4,450)$
    c. $4,450 - 0.1(4,450)$

P 7. Pantsuit Heaven took 30% off everything in stock last weekend. Which of the following expressions shows the amount of money Naomi paid for a pantsuit that was originally priced at $59.99?

    a. $\$59.99(0.30)$ **d.** $\$59.99 - 0.30(\$59.99)$
    b. $\$59.99 + 0.30(\$59.99)$ **e.** $\$59.99/30$
    c. $\$59.99 - .30$

P 8. Bill wants to buy the least expensive cereal he can find. Based on the information below, which should he buy?

Cereal A: $1.98, minus a 30¢ coupon
Cereal B: $2.09, minus a 35¢ coupon
Cereal C: $2.20, minus a 10% store discount
Cereal D: $1.69
Cereal E: $2.25, including a 35¢ discount on a box of doughnuts

    a. A   **b.** B   **c.** C   **d.** D   **e.** E

P 9. A new can of roach killer is advertised as containing 25% more spray for the regular price of $3.39. If the previous can contained 28 ounces of spray, how many ounces does the new can hold?

    a. 4   **b.** 4.24   **c.** 7   **d.** 35   **e.** 53

M 10. How many 6-inch-long pieces of wood are needed to cover 1 yard of garden border?

    a. $\frac{1}{2}$ **b.** 5   **c.** 6   **d.** 7   **e.** not enough information given

CHECKING THE ANSWER

Did You Answer the Question?

By 2:00 P.M., each worker at Smart's Dry Cleaners had pressed 120 shirts. To finish out the day, the 3 workers divided equally the remaining 60 shirts. By the end of the day, how many shirts did each worker press?

a. 180   **b.** 140   **c.** 60   **d.** 20   **e.** not enough information given

How many operations does it take to solve this problem? Do you think 20 is the correct answer? Or 140?

If you found the answer of this problem to be 20, you have made a common mistake. You will see your error and help avoid similar mistakes in the future by checking to see if you *answered the question asked.*

> **TIP**
> Rereading the question is very important in multistep problems like the one here. The solution for the first step is often listed as an answer choice for the problem, so watch out!

STEP 1	Find how many shirts each worker pressed after 2:00 P.M.	$60 \div 3 = 20$ shirts
STEP 2	Reread the question.	The question asks for the total that each worker pressed—not the amount after 2:00 P.M.
STEP 3	Add the two amounts.	120 shirts (by 2:00 P.M.) + 20 shirts (after 2:00 P.M.) 140 shirts per worker The correct answer is **b. 140.**

Answer the Question Asked

Let's take a look at another problem. Pay special attention to what you are being asked to find.

Jacqueline bought a leather jacket during a $\frac{1}{3}$-off sale at a department store. The jacket was originally priced at $270. How much did Jacqueline save by buying the jacket on sale?

a. $360   **b.** $270   **c.** $180   **d.** $90   **e.** $9

What is the answer to this problem?
Does your solution answer the question asked?

STEP 1   To get the answer, find $\frac{1}{3}$ of $270.   *$270 ÷ 3 = $90*

STEP 2   If you think you should subtract $90 (amount saved) from $270 (original price), REREAD THE QUESTION.

The question asks for the amount saved—NOT the amount paid!
*The correct answer is **d. $90**.*

First solve each problem, paying close attention to what you are being asked to find. Then look at the *incorrect* answer given below "your solution" and write down *what question* this incorrect solution is answering. This will help you focus on answering the question that was asked.

> **TIP**
> When you are solving a problem, use **labels** whenever you can. Then compare the label in your answer to the label asked for in the question.

<u>EXAMPLE</u>   In 1998, Caroline's neighborhood association had 130 members. Over the next two years, that number increased by 20%. How many members did the association have in 2000?

   a. 26   **b.** 104   **c.** 146   **d.** 156   **e.** not enough information given

Your solution: **STEP 1** $130 \times 0.20 = 26$ members

   STEP 2 $130 + 26 = 156$ members

The correct answer is **d. 156.**

Choice **a.** is incorrect because it answers the question: *By how many members did the association increase?*

WN **1.** A land-developing company owned 1,350 acres in Red County last year. This year the company added 200 acres in March, 750 acres in June, 90 acres in September, and 110 acres in November. How many total acres were added this year?

    a. 2,500   **b.** 1,150   **c.** 1,060   **d.** 950   **e.** 200

Your solution:

Choice **a.** is incorrect because it answers the question: _____

_____

D **2.** Maureen started the week with $404.00 in the bank. During the week she wrote checks for $37.50, $18.97, and $144.00. She also deposited a tax refund of $200.00. How much did Maureen have in her account at the end of the week?

    a. $200.47   **b.** $337.53   **c.** $403.53   **d.** $604.00   **e.** not enough information given

Your solution:

Choice **a.** is incorrect because it answers the question: _____

_____

P **3.** Sixty-five percent of the inmate population at Reeves Correctional Facility is Caucasian. The number of African Americans and other minority inmates is 700. What is the total number of inmates at Reeves?

    a. 4,000   **b.** 2,000   **c.** 1,300   **d.** 70   **e.** 35

Your solution:

Choice **c.** is incorrect because it answers the question: _____

_____

P 4. As an employee of a store, Floyd gets 15% off anything he purchases. If he bought a suitcase priced at $35 and a pair of shoes priced at $40, what did Floyd actually pay for these two items (before taxes)?

    a. $75   **b.** $63.75   **c.** $28.75   **d.** $23.75   **e.** $11.25

Your solution:

Choice **e.** is incorrect because it answers the question: _____

_____

M 5. To a bucket that contained 1 quart of water, a janitor added $\frac{3}{4}$ quart of ammonia and 2 cups of liquid detergent. How many *quarts* of liquid were in the bucket?

    a. $2\frac{1}{4}$ **b.** $3\frac{3}{4}$ **c.** 5   **d.** 10   **e.** not enough information given

Your solution:

Choice **c.** is incorrect because it answers the question: _____

_____

D 6. Norris had $205.00 in his checking account. During the week he wrote checks for his electric bill of $19.25, his gas bill of $8.55, the pharmacy for $22.15 and a restaurant for $10.75. How much did Norris spend on utility bills?

    a. $27.80   **b.** $60.70   **c.** $144.30   **d.** $177.20   **e.** $49.95

Your solution:

Choice **b.** is incorrect because it answers the question: _____

_____

Is the Answer Reasonable?

Marcie's checking account balance was $890.00. She then wrote checks for $110.00 to the electric company, $42.33 to the telephone company, and $21.90 to a department store. What is her new balance?

a. $1,605.77   **b.** $1,064.23   **c.** $954.23   **d.** $715.77   **e.** $315.77

Suppose you added $110.00, $42.33, and $21.90 to $890.00 and chose b. $1,064.23 as your answer. How can you tell that you have done something wrong?

One part of checking your answer is deciding whether your solution is **reasonable**—that is, does it make sense?

Why is answer choice **b.** unreasonable? If Marcie started with $890 and wrote checks (subtracted), it doesn't make sense that her new balance is *larger* than the opening balance!

Here is the correct solution:

$890.00	beginning balance
− $110.00	check to electric company
$780.00	
− $42.33	check to phone company
$737.67	
− $21.90	check to department store
$715.77	

The correct answer is **d. $715.77.**

A quick check will tell you that $715.77 is a reasonable amount to have left in the account. You can check by rounding the numbers.

$900 − $100 − $40 − $20 = $740 estimated
rounded   ↑   ↑   ↑   ↑
from   890   110 42.33 21.90

$740 ≈ choice **d. $715.77**

Now take a look at another problem on the top of page 149.

Some people believe that if you double a child's height at age two you will have the height of that child as an adult. According to this rule, if James was 34 inches at age two, how many inches would he be as an adult?

a. 2   **b.** 17   **c.** 36   **d.** 51   **e.** 68

Suppose you divided 34 inches by 2 and chose b. 17 as your answer. Is 17 inches a reasonable height for an adult?

TIP
Once you have solved a problem, make a statement out of the question by inserting your answer. Think about your statement. Does it make sense? In the example at the right, you could see that this sentence does *not* make sense: "James, who was 34 inches at age 2, is now 17 inches tall."

Quickly compare your answer to the facts given in the problem. Seventeen inches is *not* a reasonable height for an adult—especially if he was 34 inches tall as a child.

The correct solution is found this way:

$$34 \times 2 = 68 \text{ inches}$$

age 2   doubled   adult

The correct answer is **e. 68.**

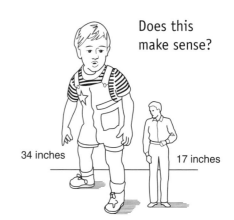

Does this make sense?

34 inches   17 inches

First solve each problem, paying close attention to what you are being asked to find. Then look at the *incorrect* answer given below "your solution" and write down why the answer choice is unreasonable.

EXAMPLE   Nancy started work at 5:00 P.M. She spent 2 hours dusting, $1\frac{1}{2}$ hours emptying wastebaskets, and 1 hour sweeping. What time did Nancy finish this work?

a. 12:30 P.M.   **b.** 1:30 P.M.   **c.** 7:30 P.M.   **d.** 9:30 P.M.   **e.** 10:00 P.M.

Your solution:

STEP 1 $2 \text{ hr} + 1\frac{1}{2} \text{ hr} + 1 \text{ hr} = 4\frac{1}{2} \text{ hr}$

STEP 2 $5:00 \text{ P.M.} + 4\frac{1}{2} \text{ hr} = 9:30 \text{ P.M.}$

The correct answer is **d. 9:30 P.M.**

Choice **a.** is unreasonable because: *It is $4\frac{1}{2}$ hours earlier, not later, than when Nancy started.*

WN 1. Della went on a new exercise program and diet. She lost 23 pounds. Before this loss she weighed 157 pounds. What is Della's new weight, in pounds?

    a. 200   **b.** 180   **c.** 155   **d.** 134   **e.** not enough information given

Your solution:

Choice **b.** is unreasonable because: _____

_____

WN 2. Last week Sharen worked the hours shown in the chart. What was the *average* number of hours she worked daily over this 7-day period?

Day	Hours
Monday	8
Tuesday	6
Wednesday	3
Thursday	8
Friday	5
Saturday	3
Sunday	2

    a. 40   **b.** 35   **c.** 7   **d.** 5   **e.** 2

Your solution:

Choice **b.** is unreasonable because: _____

_____

WN 3. Tim drove 60 miles per hour for 4 hours and 50 miles per hour for 5 hours. How many miles did he travel in all?

    a. 9   **b.** 10   **c.** 15   **d.** 25   **e.** 490

Your solution:

Choice **d.** is unreasonable because: _____

_____

Checking Your Computation

Mrs. Ruben paid $1.41 for 3 bottles of vinegar and $5.07 for 3 pounds of meat. How much would she pay for one bottle of vinegar and one pound of meat?

a. $0.47   **b.** $1.69   **c.** $2.16   **d.** $6.48   **e.** $7.92

What operations do you use to find the solution?
How do you check your work?

Suppose you are sure you have answered the question being asked, and you are certain that your answer is reasonable. However, you can't find your solution listed in the choice of answers.

One final step in checking your answer is *making sure you have done the computation correctly.* Follow the examples below to see how to do this check.

Perform each operation carefully.

STEP 1

$$\begin{array}{r} \$0.47 \text{ per bottle} \\ \hline 3\,)\overline{\$1.41} \end{array}$$

Check:
$$\begin{array}{r} \$0.47 \\ \times\ \ 3 \\ \hline \$1.41 \end{array}$$

STEP 2

$$\begin{array}{r} \$1.69 \text{ per pound} \\ \hline 3\,)\overline{\$5.07} \end{array}$$

Check:
$$\begin{array}{r} \$1.69 \\ \times\ \ 3 \\ \hline \$5.07 \end{array}$$

STEP 3

$$\begin{array}{r} \$0.47 \\ +\ \$1.69 \\ \hline \mathbf{\$2.16}\ \textbf{total} \end{array}$$

Check:
$$\begin{array}{r} \$2.16 \\ -\ \$1.69 \\ \hline \$0.47 \end{array}$$

You're right if you said the answer is **c. $2.16.**

....

Did You Remember . . . ?
- To check an addition problem, you subtract.
- To check a subtraction problem, you add.
- To check a division problem, you multiply.
- To check a multiplication problem, you divide.

....

All of these problems are wrong! Find the computation errors by using the checkup steps given on page 151.

EXAMPLE

$$\begin{array}{r} 345 \\ + 428 \\ \hline 763 \end{array}$$

Check:
$$\begin{array}{r} 763 \\ - 428 \\ \hline 335 \end{array}$$

Corrected:
$$\begin{array}{r} 345 \\ + 428 \\ \hline 773 \end{array}$$

Recheck:
$$\begin{array}{r} 773 \\ - 428 \\ \hline 345 \end{array}$$

1.
$$5\overline{)975} \quad 185$$

Check:

Corrected:

Recheck:

2.
$$\begin{array}{r} 105 \\ \times \quad 30 \\ \hline 31,500 \end{array}$$

Check:

Corrected:

Recheck:

3.
$$\begin{array}{r} 1,009 \\ - \quad 437 \\ \hline 562 \end{array}$$

Check:

Corrected:

Recheck:

4.
$$23\overline{)11,201} \quad 481$$

Check:

Corrected:

Recheck:

Mixed Review

Choose the correct answers for the following problems.

WN **1.** A customer put $25 down for a coat he put on layaway. He agreed to pay off the balance in 3 installments over the next 3 months. What is the total cost of the coat?

 a. $25 **b.** $75 **c.** $100 **d.** $125 **e.** not enough information given

WN **2.** A 2-foot square carton holds 24 cans of pie filling. If a packer has 640 cans to pack, how many of these cartons can he fill completely?

 a. 24 **b.** 25 **c.** 26 **d.** 27 **e.** 15,360

WN **3.** According to the graph, which of the states shown had the highest per-capita personal income in 1998?

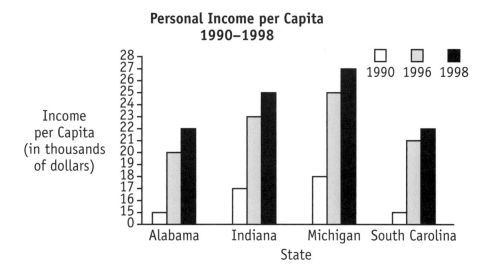

 a. Alabama **d.** South Carolina
 b. Indiana **e.** not enough information given
 c. Michigan

WN **4.** A schoolteacher bought 6 packages of a dozen pencils each and distributed them to her 18 students. Which expression shows the number of pencils received by each student?

 a. $6 \div 18$ **d.** $18(6 \times 12)$
 b. $18 \div 6$ **e.** $\dfrac{6 \times 12}{18}$
 c. $12(18 \div 6)$

WN 5. For every 2 job candidates that Ned turns away he usually hires 3. If, in a given month, Ned hires 12 people, how many candidates did he turn away?

    a. 1   **b.** 2   **c.** 4   **d.** 8   **e.** 24

D 6. Mr. Tyson cuts a 14-inch piece of wire into lengths of 0.35 inches. How many pieces of wire does he get?

    a. 4.9   **b.** 40   **c.** 80   **d.** 400   **e.** 540

F 7. McCarthy's Deli has a $\frac{1}{4}$-off sale on all prepared salads. How much does Hannah save by buying 2 pounds of potato salad originally priced at $1.50 per pound?

    a. $0.37   **b.** $0.75   **c.** $1.13   **d.** $2.00   **e.** $2.25

M 8. The chart below shows the regular hours worked by Mike's construction crew. What more do you need to know to find out how much money Mike paid for the payroll that day?

Crew	Hours Worked
Henry Elias	8:00 A.M.– 5:00 P.M.
Yolanda Grace	7:00 A.M.– 4:00 P.M.
Mitch Stewart	7:00 A.M.– 4:00 P.M.
Ronald Jones	7:00 A.M.– 4:00 P.M.

    a. the total hours worked by the crew
    b. the number of overtime hours put in
    c. Mike's weekly payroll
    d. the amount each worker is paid per hour
    e. Mike's expenses other than payroll

Problems 9 and 10 refer to the following information.

When she shops, Ellen buys only the most economical cleaning supplies, frozen foods, and canned goods. Then she buys 3 pounds of whatever meat is on sale.

Last week the supermarket had Luxury dish soap at $1.60 and Most dish soap at $1.92. In addition, 16 ounces of Nature frozen corn was on sale for $2.03, while Better frozen corn was priced at $2.29 per pound.

D 9. Before sales tax, approximately how much did Ellen pay for one bottle of dish soap and one bag of frozen corn?

    a. $8.00   **b.** $4.50   **c.** $3.50   **d.** $43.00   **e.** not enough information given

D 10. If round steak was on sale for $2.49 per pound, about how much did Ellen pay for meat this week?

    a. $2.50   **b.** $5.00   **c.** $7.50   **d.** $10.00   **e.** not enough information given

D 11. Ellen's cousin, Nita, bought the more expensive dish soap and frozen corn. She also bought 2 pounds of hamburger for $1.59 per pound. About how much did Nita spend before sales tax?

    a. $4.00   **b.** $3.00   **c.** $6.00   **d.** $10.00   **e.** $7.50

USING CHARTS, GRAPHS, AND DRAWINGS

Reading a Picture

According to the graph, what is the average price of a 32-ounce box of oat bran cereal?

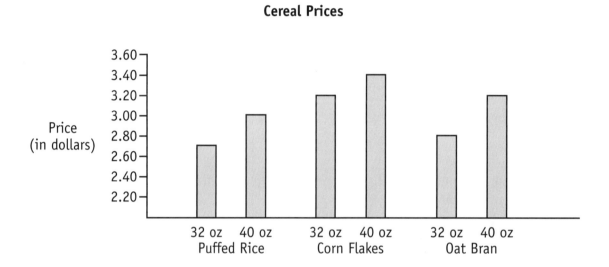

Cereal Prices

Could you answer the question without looking at the graph? How do you go about finding the right information on the graph?

Graphs, charts, tables, and drawings are all ways to show a picture of some information. Rather than use long lists of words and numbers, people can use these tools to make information easy to understand and compare. Many word problems use graphs, charts, and pictures because they are real-life examples of math in action.

To answer the problem above,

- **Read the accompanying graph.** The problem states "according to the graph." You are being asked to use information from the drawing—not from your own experience.

- **Find the oat bran category of cereals listed.** Oat bran is represented by the last two bars on the graph.

- **Find the correct box size in the oat bran category.** The 32-ounce size is the first bar for oat bran.

Now that you have found the spot where your information is located,

- **Scan across the graph** to the vertical axis where the actual prices are listed, as shown below.

Cereal Prices

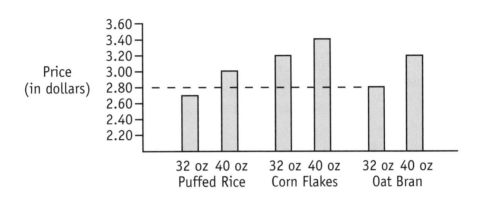

- **Decide what the price is.** In this problem, the 32-ounce oat bran cereal bar rises exactly to the **$2.80** mark.

Now take a look at another problem, based on the same graph.

According to the graph, what was the average price *per ounce* of oat bran cereal in the 40-ounce package?

a. $3.20   **b.** $1.20   **c.** $0.80   **d.** $0.08   **e.** $0.02

How is this problem different from the one on page 156? What are the operations needed to solve it?

To solve this problem, you'll need to do all the same operations you did for the last problem. However, there is another step.

Once you have found the package price ($3.20), you need to find the unit price (price per ounce).

$$\begin{array}{r} \$0.08 \leftarrow \text{price per ounce} \\ 40\overline{)\$3.20} \end{array}$$
↑   ↑
number package
  of   price
ounces

The correct answer is **d. $0.08.**

> **TIP**
> Some word problems will ask you only to find a value on a graph or chart. Other word problems will require you to find a value and do some computation with this value.

Using Charts

Take a look at a problem based on a chart of information.

Information from a magazine questionnaire was placed in the chart at the right. If 1,000 people responded, how many of them pay under $100 per week for child care?

a. 10   **b.** 50   **c.** 100   **d.** 150   **e.** 1,000

What are the steps involved in getting the correct solution? Do you need to just locate information, or do you need to compute as well?

What People Pay for Child Care	
Under $100	10%
$100–$150	52%
$151–$250	33%
Over $250	5%

Once you have found the category for "Under $100," read across and find the correct percent. The chart shows that 10% of the people paid under $100 per week.

10% of 1,000 =

$0.10 \times 1,000 = 100$ (or $\frac{1}{10} \times 1,000 = 100$) $\longrightarrow$

The correct answer is **c. 100.**

What People Pay for Child Care	
Under $100	10%
$100–$150	52%
$151–$250	33%
Over $250	5%

....

Solve the following problems using the information provided. Problems 1–3 refer to the graph.

Percent of Population Living Outside Metropolitan Areas

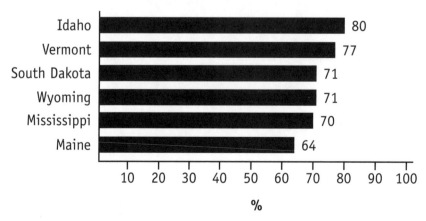

	%
Idaho	80
Vermont	77
South Dakota	71
Wyoming	71
Mississippi	70
Maine	64

WN **1.** What percent of the population in Wyoming lives outside the metropolitan areas?

    a. 29   **b.** 64   **c.** 70   **d.** 71   **e.** 80

WN 2. What percent of the population in Idaho lives inside a metropolitan area?

    a. 20   **b.** 23   **c.** 24   **d.** 77   **e.** 80

P 3. The population of Mississippi is approximately 2,600,000. Which expression shows the number of people living outside metropolitan areas in Mississippi?

    a. $70 \times 2{,}600{,}000$
    b. $70{,}000 \times 2{,}600{,}000$
    c. $0.70 \times 2{,}600{,}000$
    d. $0.70 \div 2{,}600{,}000$
    e. $2{,}600{,}000 \div 0.70$

 Questions 4–6 refer to the graph below.

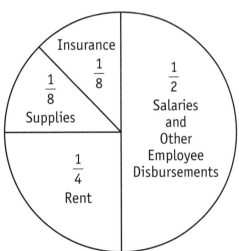

**Comco Corporation
Year 2000 Budget**

Insurance $\frac{1}{8}$

$\frac{1}{8}$ Supplies

$\frac{1}{4}$ Rent

$\frac{1}{2}$ Salaries and Other Employee Disbursements

F 4. What fraction of the budget for Comco Corporation is *not* spent on rent?

    a. $\frac{1}{8}$ **b.** $\frac{3}{8}$ **c.** $\frac{1}{2}$ **d.** $\frac{5}{8}$ **e.** $\frac{3}{4}$

F 5. How many times more money does Comco spend on rent and insurance than it does on supplies?

    a. $\frac{1}{8}$ **b.** 2   **c.** 3   **d.** 4   **e.** 5

F 6. If the entire budget for Comco in 2000 was $800,000, how much money did the company spend on insurance?

    a. $\frac{1}{8}$ **b.** $7,000   **c.** $10,000   **d.** $100,000   **e.** $700,000

Reading Between the Lines on Graphs

According to the graph, what was the approximate number of residents in the Sanford Community Housing Project in 1999?

a. 15 **b.** 1,500 **c.** 1,700 **d.** 2,000 **e.** not enough information given

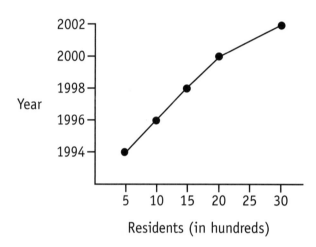

Residential Population Sanford Community Housing Project

Is this a harder problem to solve than those in the last lesson? Why?

You need to find 1999 and look across to find the correct number of residents. The tricky part of this problem is that you do *not* see 1999 clearly labeled on the graph.

To solve this type of graph problem, you'll need to **estimate.**

The answer is between 15 hundred (1,500) and 20 hundred (2,000). The value you are trying to find is a little closer to 15 than 20. Therefore, a good estimate would be **c. 1,700 residents.**

The next exercise will give you more practice in estimating with graphs.

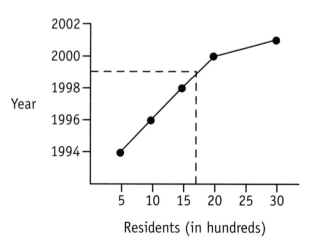

Residential Population Sanford Community Housing Project

Solve the following problems using the graphs provided.

Problems 1–3 refer to the graph below.

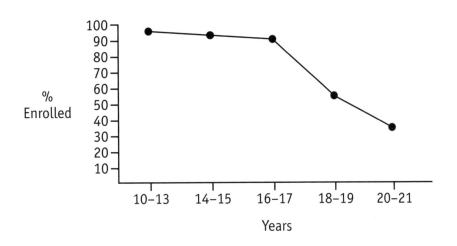

United States School Enrollment

WN 1. What age group (between 10 and 21 years) had the smallest percent enrolled in school?

   a. 10–13   **b.** 14–15   **c.** 16–17   **d.** 18–19   **e.** 20–21

WN 2. Approximately what percent of all 14- and 15-year-olds were enrolled in school?

   a. 9   **b.** 31   **c.** 51   **d.** 91   **e.** 99

P 3. If school enrollment in the town of Westville is typical of school enrollment across the United States, approximately how many of Westville's 5,500 18- and 19-year-olds are enrolled in school?

   a. 55   **b.** 100   **c.** 3,025   **d.** 4,000   **e.** 5,500

Problems 4–6 refer to the following graph.

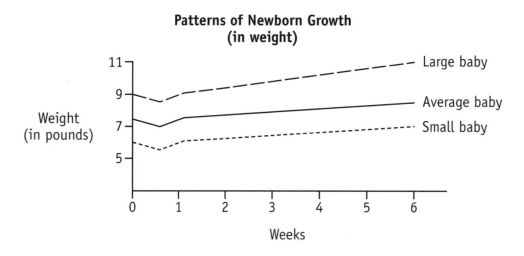

WN 4. According to the graph, most newborns

    a. gain weight during their first week, then lose weight
    b. lose weight during their first week, then gain weight
    c. lose weight during their first 6 weeks
    d. stay the same weight during their first 6 weeks
    e. weigh less than 6 pounds at birth

WN 5. By the age of 6 weeks, what is the approximate difference in pounds between a large baby and a small baby?

    a. 0   **b.** 4   **c.** 6   **d.** 8   **e.** 11

WN 6. At the end of the first week, what is the approximate difference in pounds between an average baby and a small baby?

    a. 2   **b.** 4   **c.** 6   **d.** 8   **e.** 9

Reading Graphs and Charts Carefully

According to the graph, how many offices did the Handle Company clean in 1999?

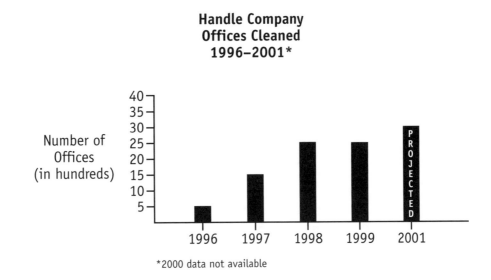

**Handle Company
Offices Cleaned
1996–2001***

Number of
Offices
(in hundreds)

*2000 data not available

a. 5   **b.** 15   **c.** 25   **d.** 2,500   **e.** 3,500

John quickly found the 1999 bar on the graph and read across to the value of 25. He chose **c. 25** as his answer.

What mistake did John make?

When you solve a problem based on a graph, you must be careful to read *all* of the information given. Although John correctly found 25 on the graph, he failed to see the label *(in hundreds)* included on the side of the graph. The correct answer to this problem is **d. 2,500**.
$(25 \times 100 = 2,500)$

Try another problem.

According to the graph, how much will Nancy weigh in March of 1999?

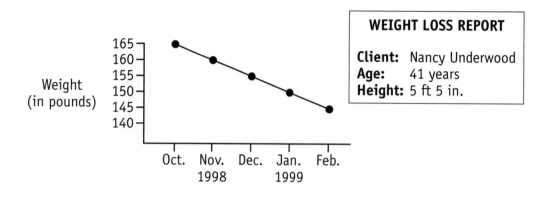

a. 145   **b.** 140   **c.** 135   **d.** 130   **e.** not enough information given

Pauline looked at the graph and noticed that Nancy was losing weight at a rate of 5 pounds per month. She subtracted 5 pounds from 145 pounds, Nancy's weight in February 1999. She chose **b. 140** as her answer.

What mistake did Pauline make?

Pauline made one major mistake. The graph does not provide *any* information about March of 1999. She cannot assume that Nancy will weigh 5 pounds less in March than in February. (Perhaps Nancy started a different diet and lost 8 pounds instead; or perhaps 145 pounds was her desired weight.) The correct answer to this problem is **e. not enough information given.**

TIP
When using a graph, use *all* the labels and information provided. However, don't make assumptions or predictions unless you are specifically asked to.

Use the information provided to answer the following problems.

 Problems 1–3 refer to the following graph.

Fat Content of
Common Food Items

Lowfat Yogurt
(8 oz)   ● ● ● ●   ● = 1 gram fat

2% Milk
(1 cup)   ● ● ● ● ●

Roast Chicken, no skin
(3 oz)   ● ● ● ● ● ●

Fried Beef Liver
(3 oz)   ● ● ● ● ● ● ● ●

Ground Beef
(3 oz)   ● ● ● ● ● ● ● ● ● ● ● ● ● ● ● ● ●

WN 1. How many fewer grams of fat are in 3 ounces of fried beef liver
than in 3 ounces of ground beef?

a. 8   **b.** 9   **c.** 17   **d.** 24   **e.** not enough information given

WN 2. Belinda drinks 3 cups of 2% milk every day. How many grams of
fat does she consume in this milk?

a. 5   **b.** 10   **c.** 15   **d.** 20   **e.** not enough information given

WN 3. Quentin ate 6 ounces of roast chicken with skin. How many grams
of fat did the chicken contain?

a. 3   **b.** 4   **c.** 5   **d.** 6   **e.** not enough information given

Problems 4–7 refer to the chart below.

Indianapolis 500 Auto Race Winners			
Year	Driver	Time	MPH (miles per hour)
1911	Ray Harroun	6 hr 42 min 8 sec	74.59
1931	Louis Schneider	5 hr 10 min 27.93 sec	96.629
1951	Lee Wallard	3 hr 57 min 38.05 sec	126.244
1971	Al Unser	3 hr 10 min 11.56 sec	157.735
1987	Al Unser	3 hr 4 min 59.147 sec	162.175
1998	Eddie Cheever	3 hr 26 min 40.524 sec	145.155

D 4. How many miles per hour did the winning car travel in 1987?

    a. 59.147   **b.** 74.59   **c.** 162   **d.** 162.175   **e.** not enough information given

D 5. To win the race, approximately how many times longer did Ray Harroun take than Al Unser in 1971?

    a. $\frac{1}{2}$ **b.** 2   **c.** 4   **d.** 5   **e.** not enough information given

D 6. The slowest car in the 1971 Indianapolis 500 traveled at how many miles per hour?

    a. 74.59   **b.** 96.629   **c.** 126.244   **d.** 157.735   **e.** not enough information given

D 7. About how many miles per hour faster did Al Unser travel in the 1987 race than in the 1971 race?

    a. 3   **b.** 4   **c.** 6   **d.** 10   **e.** not enough information given

Mixed Review

Choose the correct answers for the following problems.

 Problems 1–4 refer to the graph below.

WN 1. Hannah drew this graph to show how she spends her day. How many hours per day does Hannah watch television?

    a. 1   **b.** 2   **c.** 4   **d.** 5   **e.** not enough information given

F 2. Hannah feels she needs more sleep at night. If she were able to cut her cleaning and laundry time by $\frac{1}{3}$, how many total hours could she then spend sleeping?

    a. 1   **b.** 6   **c.** $6\frac{1}{3}$ **d.** 7   **e.** 8

F 3. What fraction of her day does Hannah spend preparing meals?

    a. $\frac{3}{1}$ **b.** $\frac{1}{3}$ **c.** $\frac{1}{4}$ **d.** $\frac{1}{6}$ **e.** $\frac{1}{8}$

P 4. What percent of her day does Hannah spend sleeping?

    a. 6   **b.** 12   **c.** 20   **d.** 24   **e.** 25

WN 5. Floyd planned to make a ceramic decoration for each member of his family. Each decoration takes 1 hour to design, 2 hours to mold, and 1 hour to glaze. What more do you need to know to find out how long it will take Floyd to make all the decorations?

    a. the size of each decoration
    b. the amount Floyd will spend on materials
    c. the total hours that each decoration takes to make
    d. the number of people in Floyd's family
    e. the number of decorations Floyd can make in a day

Problems 6–8 refer to the following information.

While shopping, Jim saw this sign. He ordered 2 pounds of coleslaw, $1\frac{1}{2}$ pounds of roast beef, 3 pounds of bologna, and 1 pound of American cheese. He then went to the produce section, where he chose some carrots, greens, and tomatoes. On his way to the checkout, Jim also chose a vegetable peeler for $3.20. He paid for all purchases with a $50.00 bill.

TODAY'S SPECIALS	
Roast Beef	$4.98/lb
American Cheese	$2.28/lb
Bologna	$2.99/lb
Turkey Breast	$3.99/lb
Potato Salad	$2.10/lb
Cole Slaw	$1.90/lb

D 6. Jim's total purchases, including tax, came to $47.12. How much change should he receive?

    a. $1.10   **b.** $2.88   **c.** $31.90   **d.** $97.12   **e.** not enough information given

F 7. In addition to receiving the special price on American cheese, Jim had a $0.20-off coupon for this item. Which expression shows what he would pay for the cheese?

    a. $2.28 – $0.20   **d.** 4($2.28 – $0.20)
    b. $\frac{1}{4}$($2.28) – $0.20   **e.** $2.28 + $\frac{1}{4}$($0.20)
    c. $\frac{1}{4}$($2.28 + $0.20)

P 8. The vegetable peeler was subject to a 5% tax. What was the price of the peeler including tax?

    a. $0.16   **b.** $3.20   **c.** $3.36   **d.** $4.20   **e.** not enough information given

WORKING GEOMETRY WORD PROBLEMS

Finding Information on Drawings

The drawing shows the cushion that Joseph needs to reupholster. What is the distance, in inches, around the cushion?

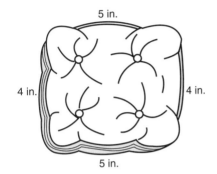

How do you find the correct answer to this problem?

You probably know that by adding up the lengths of each of the sides of the figure you'll get the total distance around, called the **perimeter** of the figure.

$4 + 5 + 4 + 5 = $ **18 inches**

To get the correct answer, you simply use the numbers written on the drawing—these numbers represent length.

Now look at another way that information can be represented on a drawing.

According to the drawing, how many feet of barbed wire will a farmer need to surround the rectangular field?

a. 8   **b.** 10   **c.** 18   **d.** 36   **e.** not enough information given

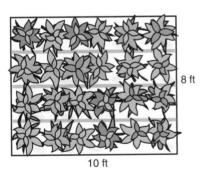

What is the length of each side of the rectangle? How did you know?

At first it may seem that some information is left off the drawing. There are four sides to the figure, but only two lengths are labeled. Is the answer **c. 18?** Or is it **e. not enough information given?**

To answer correctly, you need to reread the problem. It states that the figure in the drawing is a *rectangle.* Your knowledge of geometry should remind you that in a rectangle the opposite sides are equal in length.

$$(2 \times 8) + (2 \times 10) = 16 + 20 = 36$$
$$\uparrow \qquad \uparrow$$
2 widths   2 lengths

The correct answer is **d. 36.**

> **TIP**
> If you are having trouble with a geometry problem, reread the problem AND look at the drawing again. You may have overlooked some information.

 Solve the following word problems using the information included in the drawings.

GE **1.** A sketch of Mrs. Fein's floor plan is at the right. Her bathroom is square in shape, and both the kitchen and family room are rectangles. How many feet of baseboard does Mrs. Fein need to buy to go around the floors of the three rooms (without subtracting for doors)?

    a. 32   **d.** 86
    b. 54   **e.** 142
    c. 56

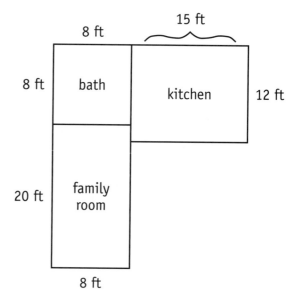

GE 2. What is the distance, in centimeters, around the figure at the right?

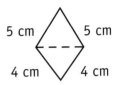

    a. 23   **b.** 22   **c.** 20   **d.** 18   **e.** not enough information given

GE 3. How many bricks would a contractor need to buy to surround the yard pictured at the right?

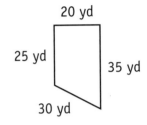

    a. 25   **b.** 35   **c.** 110   **d.** 220   **e.** not enough information given

GE 4. If fabric costs $4.99 per square yard, approximately how much will Nina need to pay to cover just the front of the three frames shown below? (**Remember:** Area = length × width.)

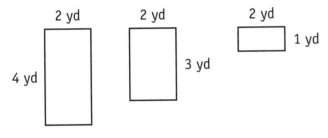

    a. $5   **b.** $6   **c.** $16   **d.** $80   **e.** $90

GE 5. A teacher cut out 20 triangles from a piece of posterboard. Each triangle had three equal sides as shown at the right. What was the perimeter, in feet, of each triangle?

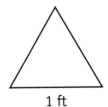

    a. 1   **b.** 3   **c.** 20   **d.** 60   **e.** not enough information given

GE **6.** How many yards of trim does Mr. Roscoe need to completely cover the perimeter of the rectangular window at the right?

    a. 2   **b.** 4   **c.** 6   **d.** 8   **e.** not enough information given

2 yd

GE **7.** Harry is ordering tile for the bathroom at the right. How many square feet does he need to order?

    a. 14   **b.** 28   **c.** 48   **d.** 22   **e.** 20

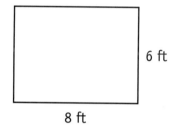

6 ft

8 ft

GE **8.** Marissa wants to put a decorative border around her bedroom and her sitting room. How many feet of border should she buy (without subtracting for the doors)?

    a. 88   **b.** 244   **c.** 44   **d.** 66   **e.** 56

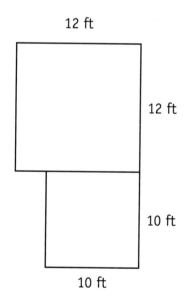

12 ft

12 ft

10 ft

10 ft

Let Formulas Work for You

To figure out how much plastic he needs to make a new kite, Ralph needs to know the surface area of this wooden frame. How many square feet of plastic will he need to cover the frame?

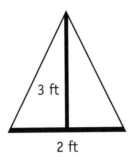

What is the shape of the kite Ralph is making?
What else do you need to know to answer this problem?

You probably can see that the kite pictured above is a **triangle** with a base of 2 feet and a height of 3 feet. To figure out how much plastic Ralph needs, you need to determine the **area** of the kite.

The AREA of a triangle $= \frac{1}{2}$ of the **base** times the height.

The formula is written as $A = \frac{1}{2}bh$.

The equation above ($A = \frac{1}{2}bh$) is called a **formula,** and, like other formulas, it is extremely useful in solving geometry problems.

> **TIP**
> Multiplication is indicated in several ways in these formulas:
>
> - a number and a letter—$4s$ means $4 \times$ side
>
> - two letters—lw means length $\times$ width
>
> - squared or cubed—s^2 means side $\times$ side, and s^3 means side $\times$ side $\times$ side

To use a formula,

STEP 1   First write the formula you need to use.

For the problem, you'll use

$A = \frac{1}{2}bh$

STEP 2   Then **substitute** any values that you know from the problem.

You know that the height is 3 and the base is 2.

$A = \frac{1}{2}(2 \times 3)$

base   height

STEP 3   Now solve the equation.

$A = \frac{1}{2}(2 \times 3)$

$A = \frac{1}{2}(6)$ *or* **3 sq ft**

Remember that area is always expressed in square units.

(**Note:** To refresh your memory about solving equations, turn back to page 97, where you first learned the rules.)

For Your Reference
A complete list of formulas is on page 209. You may use this list for reference as you work through this chapter.

Solve the following problems. First write the formula you will use. (If you need to, use the formulas on page 209.) Then solve the problem.

EXAMPLE   Sally needs to surround a square piece of wood with a plastic border. How many inches of plastic will she need for the perimeter?

Formula: $P = 4s$

Answer: $P = 4s$
$P = 4 \times 8'' \leftarrow 1$ side
$P = 32''$

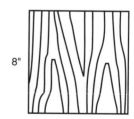
8"

GE 1. An artist begins a sculpture by forming circumferences for three wire circles of the size shown below. Which expression shows the number of inches of wire he will need?

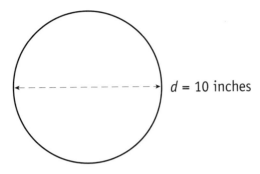

d = 10 inches

   a. 3.14×10

   b. $3(3.14 \times 10)$

   c. $3.14 \times 10 \times 10$

   d. $3(10 \times 10 \times 3.14)$

   e. not enough information given

Formula:

Answer:

GE 2. Using the measurements shown below, how many *yards* of fencing will a dog owner need to surround the perimeter of a dog pen?

12 ft

9 ft

15 ft

   a. 54   **b.** 36   **c.** 6   **d.** 12   **e.** not enough information given

Formula:

Answer:

GE **3.** What is the volume of cement, in cubic centimeters, that can be held in the cube below?

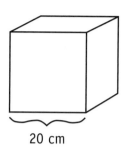

20 cm

a. 80   **b.** 400   **c.** 8,000   **d.** 160,000   **e.** not enough information given

Formula:

Answer:

GE **4.** Glowing Floor Company charges $0.50 per square foot to clean and wax wood floor surfaces. Which expression shows the amount the company would charge to clean and wax the entire wood floor shown below?

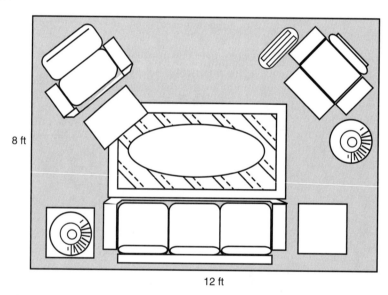

8 ft

12 ft

a. $0.50(8 + 12)$
b. $\frac{1}{2}(0.50 \times 8 \times 12)$
c. $0.50(8 \times 12)$

d. $0.50(8 + 8 + 12 + 12)$
e. not enough information given

Formula:

Answer:

Visualizing Geometry Problems

The canister shown at the right is filled to the top with oat flour. How many cubic inches of flour does it contain?

Look at the list of formulas on page 209. Which formula should you use to solve this problem? How did you decide?

h = 6 inches, *r* = 2.5 inches

Fortunately, you don't have to memorize hundreds of formulas and be able to recall them on a moment's notice. You can look in math or reference books to find the formulas you need. In fact, many kinds of tests include a list of formulas that you can use for handy reference.

But Remember:
Even when you have a list of formulas, you still need to decide *which* formula is the correct one to use. This is sometimes a tricky issue.

In the problem above there are some clues that tell you which formula to use.

- The shape shown in the drawing is a *cylinder.*

- The problem asks for the number of *cubic* inches.

Only the formula for the *volume of a cylinder* will work in this case.

$V = \pi r^2 h$

$V = 3.14 \times (2.5)^2 \times 6$

$V = 3.14 \times 6.25 \times 6$

$V = \textbf{117.75 cubic inches}$

Clues for Deciding Formulas

1. **What kinds of units are being used?**

   - *Cubic* units are used in **volume** problems.

   - *Square* units are used in **area** problems.

   - *Regular* units are used in **perimeter** problems.

2. **What is the shape of the figure in question?**

   - A *flat* figure means **area** or **perimeter** (or **circumference** of a circle).

   - A *three-dimensional* or *solid* figure means **volume**.

3. **Can you picture in your mind what is happening in the problem?**

   - A problem concerning a *surface,* such as carpeting, tile, fertilizer, fabric, or soil, most often is asking for **area.**

   - A problem concerning *distance around* an object, such as fencing, rope, or weather stripping that surrounds a shape, most often is asking for **perimeter.**

   - A problem that concerns how much a figure *holds* most often is asking for **volume.**

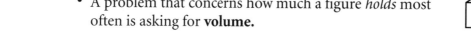

 Use the list of formulas on page 209 to solve the following problems. First circle the kind of space you will be finding. Write what clue you used. Then write the formula and solve the problem.

EXAMPLE   Max and Bertha Mahoney were responsible for roping off a children's play area at the school fair. The diameter of the circle they planned was 21 feet. How many feet of rope will they need? (Use $\frac{22}{7}$ for π.)

SPACE: perimeter/~~circumference~~ area   volume

CLUE: *roping off a circle*

SOLUTION: $C = \pi d$

$C = \frac{22}{7} \times 21$

$C = \frac{22}{7} \times \frac{21^3}{1} = 66$ feet

GE 1. How many square miles is the plot of land pictured at the right?

4 mi

2.5 mi

SPACE: perimeter/circumference   area   volume

CLUE:

SOLUTION:

GE 2. Every cubic foot of material inside the rectangular container weighs 4.5 pounds. If the box is completely filled, how many pounds of material does it contain?

SPACE: perimeter/circumference   area   volume

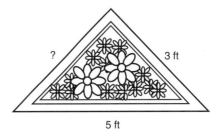

3 ft

2 ft

2.5 ft

CLUE:

SOLUTION:

GE 3. A seamstress bought 12 feet of gold braid to surround the scarf pictured at the right. How long is the third side of the scarf?

SPACE: perimeter/circumference   area   volume

?   3 ft

5 ft

CLUE:

SOLUTION:

Picturing a Geometry Problem

Peter and Lee had driven 18 miles south when they discovered they were lost. They then drove 24 miles east to get to Medfield. What is the shortest distance, in miles, between Medfield and their starting point?

How do you solve this problem? Is it a geometry problem or a simple distance problem?

If you read through this problem quickly, you might be tempted to find 42 as the answer (18 + 24 = 42). However, notice that the question does *not* ask for the total distance traveled. Instead it asks for the *shortest* distance between two points. This problem is, in fact, a geometry problem.

> **TIP**
> It is often helpful to draw a rough picture of the things taking place in a problem. This may help you to see a solution more easily.

Here are some helpful steps in solving a problem like the one above.

STEP 1   First draw a picture.

You should draw a line south and label it 18, then a line to the east and label it 24.

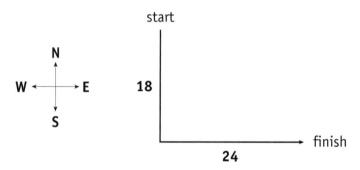

STEP 2   Next decide what the question is asking for.

*The shortest distance between **start** and **finish** would be a straight line.*

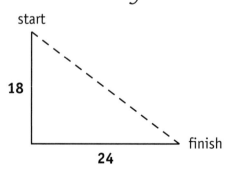

STEP 3 Once you can see the geometry problem, choose a formula that will help you.

*This is a picture of a right triangle. I can use the formula $a^2 + b^2 = c^2$ to find the length of the side (the **hypotenuse**).*

STEP 4 Substitute the values that you know, then solve.

$$c^2 = a^2 + b^2$$
$$c^2 = 18^2 + 24^2$$
$$c^2 = 324 + 576$$
$$c = \sqrt{900}$$
$$c = \textbf{30}$$

Now let's look at another geometry problem that might be hard to recognize.

How many 1-yard-square rubber pads will be needed to cover the playground area shown below?

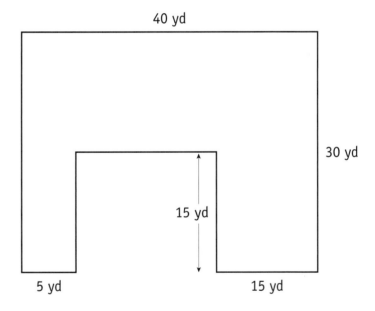

Is there a formula to determine the area of a shape like this? How can you find the area?

Don't give up if you find an unfamiliar shape like the one above. Instead, see if the shape can be broken up into smaller figures that you *do* recognize.

Here is the same shape, divided into smaller areas.

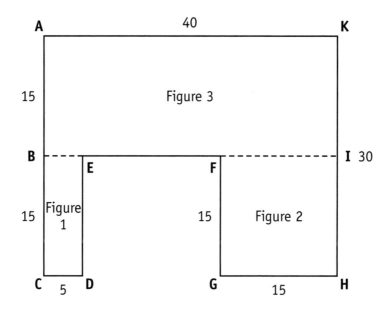

As you can see,

- Figure 1 (BCDE) is a rectangle 5 yd by 15 yd.

- Figure 2 (FGHI) is a square 15 yd by 15 yd.

- Figure 3 (ABIK) is a rectangle 15 yd by 40 yd.

To find the area of the whole figure, add the areas of the smaller figures.

Figure 1   **Figure 2**   **Figure 3**

$(5 \times 15) + (15 \times 15) + (15 \times 40) =$

75   +   225   +   600   **= 900 sq yd**

Remember:
When you have broken up the shape into smaller, more recognizable shapes, you may have to do some subtraction to find the lengths of some sides. For example, if you know that side KH is 30 yd and FG and HI are both 15 yd, you can subtract (30 − 15) to find IK.

There can be several different ways to break up a shape like the one above. Simply look for squares, rectangles, or triangles in the figure and try different ways of breaking the larger figure into smaller shapes.

Solve the following problems. Remember to draw a picture if you need help deciding what to do.

Problems 1–3 refer to the following information.

The Howes want to redo the floors in their home. The floor plan is pictured below. The carpeting they like is $8.95 per square yard installed, and the linoleum they like is $0.75 per square foot if they install it themselves.

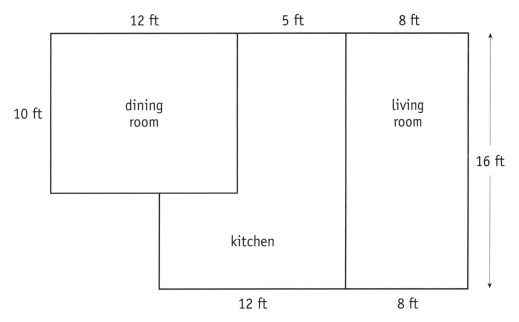

GE **1.** If the Howes decide to carpet their living room and dining room, how many square yards of carpeting do they need? (Round your answer to the nearest square yard.)

GE **2.** If they carpet both the living room and dining room, how much will the carpet cost before tax?

GE **3.** How many square feet of linoleum will the Howes need to cover the kitchen?

GE **4.** A trucker drove 160 miles north from Freeport, then drove 120 miles west to Dalton. What is the shortest distance, in miles, from Freeport to Dalton?

Mixed Review

Choose the correct answer for the following problems.

GE 1. Which expression shows the area of a circle with a radius of 5?

    a. 5×5
    b. $(3.14) \times 5 \times 2$
    c. 3.24×5
    d. 3.14×5^2
    e. not enough information given

D 2. A mother was comparing two bottles of pain relievers at the grocery store. One bottle was $2.40 for 50 tablets; the other was $3.00 for 75 tablets. What is the difference in price *per tablet?*

    a. $0.002   **b.** $0.008   **c.** $0.02   **d.** $0.20   **e.** $0.60

P 3. The Gold Company charges 7.5% monthly interest on any unpaid balance on its credit cards. In September, Mary bought a dress for $49 and a toaster oven for $79 with her credit card. When she received her bill, she paid $128. What interest does she owe?

    a. $128.00   **b.** $118.40   **c.** $9.60   **d.** $4.40   **e.** 0

Problems 4–6 refer to the following graph.

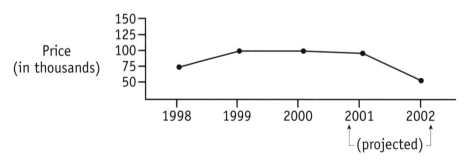

GR 4. What was the average price of a home in Case County in 1999?

    a. $75   **b.** $100   **c.** $75,000   **d.** $100,000   **e.** $175,000

GR 5. By approximately how many dollars is the price of a Case County home projected to fall between 2000 and 2001?

    a. $100,000   **b.** $90,000   **c.** $25,000   **d.** $10,000   **e.** $10

GR 6. According to the graph, what is the projected 2002 price of a Case County home?

    a. $75,000   **b.** $50,000   **c.** $25,000   **d.** $50   **e.** not enough information given

GE 7. A triangular piece of metal has a total surface area of 57 square feet. The base of the piece measures 6 feet across. Which expression shows how many feet tall the piece is?

    a. $h = 57 \times (\frac{1}{2} \times 6)$

    b. $h = 57 \div (\frac{1}{2} \times 6)$

    c. $h = 57 \div 6$

    d. $h = \frac{1}{2}(57 \times 3)$

    e. not enough information given

P 8. A bank customer earned $44.64 interest in one year for a deposit he made in a savings account. If the bank pays 8% yearly interest, how much money had the customer deposited?

    a. $3.57   **b.** $357.12   **c.** $558.00   **d.** $600.00   **e.** not enough information given

GE 9. An aluminum can holds 19.85 cubic centimeters of liquid. What more do you need to know to find the height of the can?

    a. the total weight of the filled can
    b. the weight of the empty can
    c. the volume of the can
    d. the length of a side of the can
    e. the radius of the can

F 10. Last month Marcel, an artist, sold a sculpture for $400.00, an oil painting for $797.50, and a watercolor for $297.50. Approximately what fraction of these earnings does the watercolor represent?

    a. $\frac{1}{2}$ **b.** $\frac{1}{3}$ **c.** $\frac{1}{4}$ **d.** $\frac{1}{5}$ **e.** not enough information given

Posttest A

This posttest gives you a chance to check your skill at taking a test and solving problems. Take your time and work each problem carefully. Some problems may not contain enough information to solve the problem. When you finish, check your answers and review any topics on which you need more work.

1. A builder purchased the parcel of land shown at the right. If each square yard cost $1.50, how much did he pay for the land?

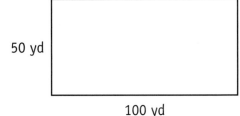

50 yd

100 yd

2. Chris is putting an above-the-ground swimming pool in his backyard. If the diameter of the pool is 20 feet, how many feet of pool siding will he need?

3. Maron is installing hardwood floors in his living room. In the first row, he is starting with the shortest piece of wood and ending with the longest. Put the following pieces of wood in the correct order, from shortest to longest.

A = 2 feet	B = 60 inches	C = 1 yard	D = $1\frac{1}{2}$ yards

4. Margie jogged 1 mile on Monday, 3 miles on Tuesday, $1\frac{1}{2}$ miles on Wednesday, and $2\frac{1}{2}$ miles on Thursday. Write an expression showing the average number of miles Margie jogged per day.

5. Thirty-five percent of families living in a small town eat red meat three times a week. The other 65 percent eat only fish and chicken. What other information is needed to find out the number of families who eat red meat?

6. Tanya worked 8 hours on Monday, 6 hours on Tuesday, and 5 hours on Wednesday. On Thursday she worked 8 hours plus 3 hours of overtime. She gets paid $7.50 per hour and additional $3.50 per hour of overtime. How much money did Tanya make in all?

7. Suzy is sewing a square quilt for her friend. For each side of the quilt, she needs 3 feet of ribbon. How much ribbon does she need in total?

 Problems 8–12 refer to the following information.

Vivian needs to make 100 copies of the company's annual report. The report is 25 pages. Vivian could have the local copy shop do the job, or she could do it herself during overtime on the office copier.

The copy shop charges 5 cents per copy. The chart at the right shows the costs if Vivian uses the office copier.

Vivian receives $13.00 per hour of overtime.

Supplies	
Toner	0.007 cents per copy
Developer	0.008 cents per copy
Paper	0.006 cents per copy
Maintenance	0.008 cents per copy

8. How much would Vivian's company pay if it had the local copy shop do the job?

9. How much does it cost to make one copy on Vivian's copier?

10. How much would it cost Vivian's company to make 100 copies of the report if Vivian did the job in 3 hours?

11. How much would Vivian's company save if Vivian did the job in 2 hours rather than the copy shop?

12. Lisa is having a dinner party at her house and is serving fish. The seafood department at the grocery store tells her to buy 2 pounds of fish for every 4 people she will be serving. If she buys 5 pounds of fish, how many people will she be serving?

13. Donna went grocery shopping and purchased two boxes of cereal at $2.99 a box, 2 pounds of bananas at 59 cents a pound, 3 pounds of potatoes at 69 cents a pound, and 1 gallon of milk at $2.69 a gallon. She gave the clerk a 20-dollar bill. How much did Donna spend on fruits and vegetables?

14. Lynda and Mark drove to New York to visit a friend. They left at 9:00 A.M. and without stopping, arrived at their friend's house at 1:00 P.M. What more do you need to know to find out how many miles per hour they traveled?

Problems 15 and 16 refer to the following graph.

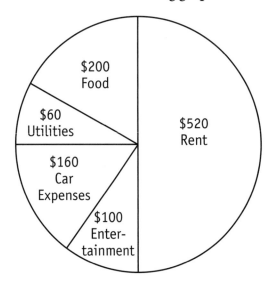

15. How much more money does Yvonne spend on rent than on food and car expenses together?

16. What fraction of the total budget is Yvonne spending on rent?

17. Happy Maids Cleaning Company cleaned one office complex and 25% of another office complex before their lunch break. There are 100 offices within each complex. Write an expression showing the amount of offices Happy Maids has cleaned.

Problems 18–20 refer to the following chart.

Top Players Points Scored in Games 1999					
Name of Player	Game 1	Game 2	Game 3	Game 4	Game 5
Mike	20	31	22	36	20
Wesley	22	32	18	38	25

18. According to the chart, how many more points did Mike score than Wesley in Game 3?

19. Wesley wants to break the 1998 season's top score of 274 points. There are five more games in the 1999 season. What is the average number of points he has to score in each game to beat the record?

20. Write an expression showing the average number of points Wesley scored per game in all five games.

21. What is the volume of sugar in the canister shown?

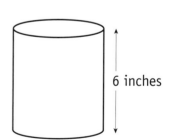

6 inches

22. If Don washes 10 windows in 57 minutes, approximately how many minutes does it take him to wash 1 window.

23. Sal is saving $30 a week to put a $250 security deposit down on an apartment. He has already saved $100. Write an expression showing how many more weeks it will take Sal to save the $250.

24. John and Kathleen are fertilizing their 13,000 square foot lawn. One bag of fertilizer covers 5,000 square feet. What is the fewest number of bags they need to purchase to fertilize their lawn?

25. Crystal typed 225 words in 5 minutes on her computer. If she continues to type at this rate, how many words can she type in 30 minutes?

26. To park in its parking garage, the museum charges $0.50 per hour for the first 3 hours and $1.50 per hour after that. If Ty parks his car for 5 hours, how much will he have to pay?

Problems 27–30 refer to the following graph.

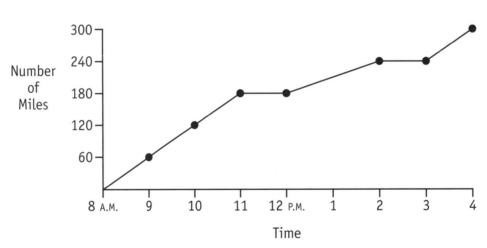

27. How many miles did the bus travel from 10:00 A.M. to 2:00 P.M.?

28. If the bus continues to travel another 60 miles, how many miles per hour will it have traveled since 1:00 P.M.?

Posttest A Prescriptions

If you missed more than one problem in any group below, review the practice pages in this book or refer to the practice pages in other materials from Contemporary Books. If you missed one problem or less in each group below, redo the problems you missed and go on to Using Number Power on page 199.

PROBLEM NUMBERS	SKILL AREA	PRACTICE PAGES
1, 6, 20, 26	multistep problems Math Exercises: Problem Solving Math Problem Solver	14–19 12–13 76–85, 150–153
21	understanding the question Math Problem Solver	25–27, 36–51 2–9, 50–59
4, 17, 20, 23	set-up questions Math Exercises: Problem Solving Math Problem Solver	44–51 10–11, 14–16 18–25, 68–75
5, 13, 14	extra information Math Exercises: Problem Solving	61–64 18–19
21, 28	not enough information Math Exercises: Problem Solving	65–68 18–19
8, 9, 10, 11	working with item sets Math Exercises: Problem Solving	71–78 20–21
12, 25	writing proportions for word problems Math Exercises: Problem Solving Math Problem Solver	101–106 22–23 168–175
22	using estimation Math Exercises: Problem Solving Math Problem Solver	121–128 8–9 12–15, 54–58, 104–109, 132–133
3	comparing and ordering	133–136
24	remainders	139–141
15, 16, 18, 19, 27	using charts, graphs, and drawings	156–166
2, 7	picturing a geometry problem Math Exercises: Geometry	177–183 6–29

Posttest B

This test has a multiple-choice format much like many standardized tests. Take your time and work each problem carefully. Circle the correct answer to each problem. You may refer to the formulas on page 209.

1. Antonio mixed 4 pints of melon, 3 pints of strawberries, and 6 pints of peaches together in a salad. He then divided the salad into 12 servings for his guests. Which expression shows the amount of salad each guest received, in pints?

   a. $\dfrac{4 \times 3 \times 6}{12}$

   b. $(12 \div 3) + (12 \div 4) + (12 \div 6)$

   c. $12 - (4 + 3 + 6)$

   d. $\dfrac{4 + 3 + 6}{6}$

   e. $\dfrac{4 + 3 + 6}{12}$

Problems 2–4 refer to the chart at the right.

2. According to the chart, what is the average number of pies rejected per day at the Masciave Bakery?

   a. 7   d. 386

   b. 49   e. 435

   c. 62

Masciave Bakery Weekly Production Report

Day	Usable Pies Produced	Number Rejected
Sunday	75	15
Monday	30	4
Tuesday	30	2
Wednesday	40	4
Thursday	60	10
Friday	100	10
Saturday	100	4

3. How many more usable pies are produced on Thursday than on Wednesday?

   a. 10   d. 100

   b. 20   e. not enough

   c. 60   information given

4. If each usable pie sells for $9.95 and each rejected pie sells at a discount for $4.00, which expression shows the amount of money the bakery takes in for pies on Monday?

   a. $34(\$9.95 + \$4.00)$ d. $(30 - 4) \times (\$9.95 + \$4.00)$

   b. $30(\$4.00) + 4(\$9.95)$ e. $(30 + 4) \times (\$9.95 - \$4.00)$

   c. $30(\$9.95) + 4(\$4.00)$

5. How many felt squares are needed to complete the quilt shown at the right?

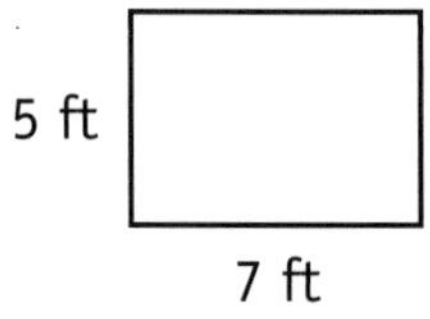

   a. 12   d. 75

   b. 24   e. not enough information given

   c. 35

5 ft

7 ft

6. If each routine examination takes 25 minutes, approximately how many of these exams can a dentist fit in if she works for 6 hours straight?

    a. 1   **b.** 6   **c.** 12   **d.** 30   **e.** not enough information given

7. Mrs. Vanderhagen canned 12 jars of plums, each containing 2 pints. She then canned five 1-pint jars of plums. Which of the following expressions will tell you how many pints she canned in all?

    a. $12 + 5$ **d.** $(2 \times 5) + 12$
    b. $2(12 + 5)$ **e.** $(12 + 2) + (5 + 1)$
    c. $(2 \times 12) + 5$

8. A medium-size U-Pick-Up vehicle can haul 500 pounds of gravel. Tim is working a job in which he needs to haul 2,400 pounds from the quarry to the worksite. If he rents the medium-size truck, what is the fewest number of trips he needs to make?

    a. 3   **b.** 4   **c.** 5   **d.** 6   **e.** not enough information given

9. Eight people can stuff 960 envelopes in 4 hours. At this rate, how many envelopes can the group stuff in 5 hours?

    a. 150   **b.** 600   **c.** 1,000   **d.** 1,200   **e.** 1,500

10. Eddie bought a half-dozen donuts at $0.49 each and two large coffees at $0.75 each. How much money did he spend in all?

    a. $2.94   **b.** $4.44   **c.** $5.88   **d.** $7.38   **e.** not enough information given

Problems 11–14 refer to the graph at the right.

Beverarges Consumed by Region
December 1999 Analysis

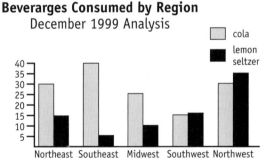

11. In what region was the consumption of cola and lemon seltzer about the same?

    a. Northeast
    b. Southeast
    c. Midwest
    d. Southwest
    e. Northwest

12. About how many more cases of cola than lemon seltzer were consumed in the Southeast in December 1999?

    a. 35   **b.** 40   **c.** 45   **d.** 35,000   **e.** 40,000

13. Half of all cola consumed in the Northwest in December 1999 was sold in metropolitan areas. How many cases of cola were sold in metropolitan areas?

    a. 15   **b.** $17\frac{1}{2}$ **c.** 15,000   **d.** 17,500   **e.** 32,500

14. During the first month of 2000, sales for diet beverages rose above sales for all colas and seltzers in the Northeast. Approximately how many cases of diet beverages were sold?

    a. 45
    b. 46
    c. 45,000
    d. 46,000
    e. not enough information given

15. How many square yards of plastic are needed to cover the rectangle shown at the right?

    a. 5   **d.** 135
    b. 9   **e.** not enough information given
    c. 45

9 ft

5 ft

16. Mrs. Soo took her Girl Scout troop on a hike through the county forest. They followed a 1-mile (5,280-foot) circular path. About how many feet did they walk to get to the very center of the circle? (**Hint:** Use 3 for the value of π when you estimate.)

    a. 31,680
    b. 1,760
    c. 880
    d. 440
    e. not enough information given

$r = \frac{1}{2}d$

17. Find the correct order in weight of the blocks shown, listing them from *heaviest* to *lightest*.

    a. B, D, A, C
    b. C, A, D, B
    c. B, C, D, A
    d. A, D, C, B
    e. D, A, B, C

A = 17 ounces   B = $1\frac{1}{2}$ pounds   C = 1 pound   D = 23 ounces

18. For his English class, John wrote a 42-page paper, a 10-page paper, and a 16-page paper. Which expression shows the average page length of John's two longest papers?

    a. $\dfrac{42+10+16}{3}$

    b. $\dfrac{42+10+16}{2}$

    c. $\dfrac{42+16}{2}$

    d. $\dfrac{42+16}{3}$

    e. $42-16$

19. Grace Anne worked 30 hours at her regular pay of $5.50 per hour. She also worked 8 hours of overtime. What more do you need to know to find out Grace Anne's rate of overtime pay?

    a. her gross pay for the 38 hours worked
    b. her total monthly hours
    c. her yearly salary before taxes
    d. the amount of her social security, state, and federal taxes
    e. her total wages for the 30 regular hours

20. Virginia mixed together 15 ounces of lemonade, 80 ounces of tonic, and 20 ounces of juice. She then poured out an 8-ounce drink. How many ounces of the mixture were left in the pitcher?

    a. 115   b. 107   c. 105   d. 95   e. 92

21. The distance around a square field is 540 yards. How many yards is one side of the field?

    a. 35   b. 40   c. 135   d. 140   e. not enough information given

Problems 22–25 refer to the following information.

Paul and Kathleen must decide where to hold their parents' anniversary party. The Knights of Columbus charges $100 per hour for rental of its hall, and food and drink can be catered or brought in. The local social club charges $75 per hour for hall rental, but food and drink must be purchased through the club. A fee of $17 per guest is charged for a buffet (including beverages).

If Paul and Kathleen decide to rent the Knights of Columbus hall, they can order party platters from their local deli. Paul's brother owns a grocery store, and they can buy beverages at cost there. The charts below show the prices they would have to pay.

Party Platters (each serves 25 people)	
Assorted cold cuts	$19.95
Breads and rolls	8.00
Vegetables and dip	12.50
Crackers and cheese	12.50

Beverages	
1 quart bottled water	$1.19
1 case (24 cans) of soda	8.80
1 quart juice	1.45

22. How much would Paul and Kathleen pay for the party, including hall rental and buffet, if they had it at the local social club for 4 hours?

    a. $75
    b. $92
    c. $300
    d. $317
    e. not enough information given

23. If they rent the Knights of Columbus hall, Paul and Kathleen have a budget of $400 for food and beverages. They will need 4 cold cut platters, 4 bread platters, 4 vegetable platters, and 4 crackers and cheese platters. How much money will they have left to spend on beverages?

    a. $200.00
    b. $211.80
    c. $188.20
    d. $88.20
    e. not enough information given

24. What would the total cost of the party be if 75 guests were invited for 3 hours at the local social club?

    a. $225
    b. $1,275
    c. $1,500
    d. $1,700
    e. not enough information given

25. Paul estimates that their guests would drink an average of $\frac{1}{2}$ quart of juice each. If 96 guests come to the party, how much will Paul need to spend on juice?

    a. $1.45
    b. $57.12
    c. $69.60
    d. $139.20
    e. $422.40

26. Taya would like to enlarge the photograph at the right. If the new length she wants is 14 inches, how many inches wide would it be?

    a. $2\frac{1}{2}$
    b. 5
    c. 10
    d. 12
    e. 14

7 in.

5 in.

27. Kerry's shopping spree resulted in a sundress for $45.00, a cookbook for $9.95, a romance novel for $5.95, and a children's book for $8.99. She received $0.11 in change. How much did Kerry spend on books alone?

    a. $70.00
    b. $69.89
    c. $25.00
    d. $24.89
    e. $24.78

28. A seamstress charges $8 per hour for hemming men's trousers. On an average day she hems 25 pairs. What more do you need to know to find the amount of money the seamstress earns on an average day?

   a. her yearly earnings
   b. the length of her lunch hour
   c. the cost of her materials
   d. the time of day she starts work
   e. the amount of time it takes to hem one pair of pants

Problems 29 and 30 refer to the circle graph at the right.

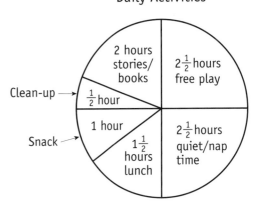

Newton Day-Care Center
Daily Activities

29. How many minutes are spent in cleanup each day at the Newton Day-Care Center?

   a. $\frac{1}{2}$ **d.** 30
   b. 5   **e.** not enough
   c. 6   information given

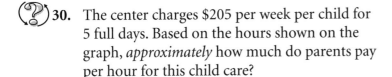

30. The center charges $205 per week per child for 5 full days. Based on the hours shown on the graph, *approximately* how much do parents pay per hour for this child care?

   a. $2.00   **d.** $40.00
   b. $4.00   **e.** not enough
   c. $8.00   information given

Posttest B Chart

If you missed more than one problem on any group below, review the practice pages for those problems. Then redo the problems you got wrong before going on to Using Number Power. If you had a passing score, redo any problem you missed and go on to Using Number Power on page 199.

PROBLEM NUMBERS	SKILL AREA	PRACTICE PAGES
10, 20	multistep problems	14–19
15	understanding the question	25–27, 36–51
1, 4, 7, 18	set-up questions	44–51
19, 27, 28	extra information	61–64
5, 14	not enough information	65–68
22, 23, 24, 25	working with item sets	71–78
9, 26	writing proportions for word problems	101–106
6, 30	using estimation	121–128
17	comparing and ordering	133–136
8	remainders	139–141
2, 3, 11, 12, 13, 29	using charts, graphs, and drawings	156–166
16, 21	picturing a geometry problem	177–183

Using
Number
Power

Smart Shopping

Have you ever thought about buying new furniture as a math problem? For example, suppose Gilberto wants to buy a new couch for his apartment. He knows the style he wants. He knows the color he wants. Does he go to the closest store and buy any couch he likes with his credit card?

If he is a smart shopper, he doesn't. Instead, he does some **research.** In other words, he looks for useful information and makes smart decisions. Here are some questions Gilberto might want to consider.

- **How big a couch can fit in the room?** Couches come in different lengths and widths.

- **How much money will my budget allow me to spend?** A beautiful leather couch might be nice, but if it takes your entire savings account just for the down payment, it's probably not a smart purchase.

- **Where can I buy this couch for the least amount of money?** Many times, you can find the same item at a discount if you are willing to look around.

- **How should I pay for the couch?** Using a credit card can sometimes be convenient, but the couch can become very expensive if you end up paying a lot of interest.

 Look at the two choices below and decide which couch would be the better purchase for each customer described on the next page.

Couch 1

- 60 inches long
- regular price $399
- 5% sales tax
- 20% discount
- no delivery charge
- free stain guard

Couch 2

- 72 inches long
- regular price $459
- no sales tax
- 25% discount
- $30 delivery charge
- matching chair can be purchased for half price
- $15 for stainguarding

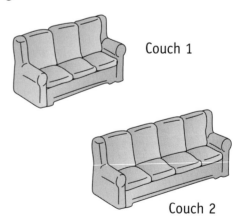

Couch 1

Couch 2

1. Customer A

   "I need a couch that can fit along a 9-foot-long wall. I can't actually purchase it until next month. My boyfriend has a truck, so I don't need to pay for delivery. I don't need any stainguarding because I am neat and don't have children or pets. Which couch is a smart purchase for me?"

   Which couch? _____   **How much will she pay?** _____

2. Customer B

   "I have saved $400 to buy a couch, and I'm ready to buy it today. I have a truck, so delivery is not a money issue. If I can find a good deal on a matching chair, I'd buy that too. My apartment is small, so I'd need a couch that is no more than 6-feet long. What do you recommend?"

   Which couch? _____   **How much will he pay?** _____

3. Customer C

   "I would like to buy a couch and have it delivered to my home. I can buy it today if I use my credit card, but my interest rate is 21% per year. Which couch should I buy if I plan to pay it off in one year?"

   Which couch? _____   **How much will he pay?** _____

4. Customers D and E

   "We'd like to buy a couch for our entertainment room. We don't care what size it is, but we don't want to pay more than $350, including delivery and stainguarding. What should we buy?"

   Which couch? _____   **How much will they pay?** _____

Ordering by Mail

Do you get catalogs and other offers to purchase items through the mail? Sometimes (but not always!) it can be an inexpensive way to shop. If you read the fine print and are sure you are getting exactly what you are paying for, mail order can make a lot of sense.

The important things to remember when you order by mail are

- Figure out ahead of time how much money you are spending. Be sure to include all costs—taxes, shipping and handling, and so on.

- Fill the form out neatly and carefully. Many orders are lost or delayed because the customer did not follow directions.

- Check your order carefully when it arrives. Be sure that you've gotten what you requested and that it is in good condition.

An order form like the one below allows you to send in camera film to be developed. The photo lab will send you your developed pictures. You can also order more film.

Payless Photo Lab

Price Guide*

Number of Exposures (pictures)	$3\frac{1}{2} \times 5\frac{1}{4}$ Prints	4×6 Prints	5×7 Prints
12–15	$4.95	$5.25	$7.25
20	$5.95	$7.95	$9.95
24–27	$6.95	$8.95	$10.95
36	$8.90	$11.95	$13.95

*Take 10% off if you are using Payless Film!
*2nd set of prints is half price!

Order Form

Name _____

Address _____

City, State, Zip _____

1. Circle print size: $3\frac{1}{2} \times 5\frac{1}{4}$ 4×6 5×7
2. How many sets of prints? 1 2
3. How many rolls are enclosed? _____
4. Cost of prints (see chart above). _____
5. Add $1.00 for shipping and handling. _____
6. New York residents add 6% sales tax. _____

TOTAL: _____

Fill in the order form for each customer below. Use a calculator to figure out total cost of the order.

1. Tino Sanchez, 11 Main Street, Milton, NY, 09000 has three rolls of film to be developed. Two are 24-exposure, and one is 36-exposure. He would like one set of prints.

> **Order Form**
>
> **Name** _____
> Address _____
> City, State, Zip _____
>
> 1. Circle print size: $3\frac{1}{2} \times 5\frac{1}{4}$ 4 × 6   5 × 7
> 2. How many sets of prints?   1   2
> 3. How many rolls are enclosed?   _____
> 4. Cost of prints (see chart above).   _____
> 5. Add $1.00 for shipping and handling. _____
> 6. New York residents add 6% sales tax. _____
>   TOTAL: _____

2. Mary Chung, 346 D Street, Apt. 4, Westend, CT, 00001 would like two sets of prints for each of her two rolls of film. Both rolls are 12-exposure. For one roll she'd like the 4 × 6 size; for the other roll she'd like 5 × 7 prints.

> **Order Form**
>
> **Name** _____
> Address _____
> City, State, Zip _____
>
> 1. Circle print size: $3\frac{1}{2} \times 5\frac{1}{4}$ 4 × 6   5 × 7
> 2. How many sets of prints?   1   2
> 3. How many rolls are enclosed?   _____
> 4. Cost of prints (see chart above).   _____
> 5. Add $1.00 for shipping and handling. _____
> 6. New York residents add 6% sales tax. _____
>   TOTAL: _____

Creating a Budget

At one time or another, most people wonder where all their money goes. It always seems as if there is more going out than coming in. Creating a detailed **budget**—a plan for how you will spend your money—is a good way to help you keep track of and even cut down on spending.

The first step in creating a budget is to understand *exactly* what you spend your money on. Large items such as rent or childcare might be obvious. But the little expenses, such as subway tokens, a fast-food meal, and laundry detergent, can add up fast.

To start budgeting, first get a small notebook that is easy to carry around with you at all times. For one complete month, write down every penny you spend. Even the 25 cents you put in the parking meter should be recorded! Here's an example of how Harry, a 23-year old single man, recorded his expenses.

Jan 1	rent	$140.00
	phone bill	$47.48
	credit card payment	$30.00
Jan 4	groceries	$22.30
Jan 5	new shoes	$42.99
Jan 12	monthly subway pass	$28.00
	groceries	$52.65
Jan 15	dinner with friends	$22.50
Jan 19	coffee and muffin	$4.40
Jan 22	library overdue fine	$0.30
	groceries	$38.70
Jan 26	dinner with friends	$24.00
Jan 27	hat and gloves	$48.00
Jan 30	cleaning service	$50.00
Jan 31	laundromat	$8.75

1. What total amount did Harry spend in January? _____
   If you can, make a list of your own expenses for one month. What is the total?

Now that you've figured out where your money goes, figure out where it comes from. Here is Harry's income. How much does he make per month?

Salary for maintenance job	*$10.80 per hour/35 hours per week*	2. _____
Weekend waiter salary	*$3.30 per hour/20 hours per week*	3. _____
Tips	*about $120 per week*	4. _____
	TOTAL MONTHLY INCOME:	5. _____

What is your total monthly income?

6. What do you think about Harry's income as compared to his expenses? Imagine that Harry has come to you for advice. He wants to put some money in a bank account each month so he can save money to go back to school. Do you think he can do this based on his current spending habits? Is Harry spending his income wisely? Where do you think he could cut back? Write a paragraph stating what advice you would give to Harry to help him save some money.

7. Now look at your own income and expenses. What can you change in order to save more money?

Using a Calculator

A calculator is a useful tool when you are solving math problems on a test. It can perform complex operations quickly, easily, and accurately. However, a calculator cannot do your thinking for you. And you must be very, very careful to enter numbers and operations carefully.

EXAMPLE   A scientist measured and recorded the following times during
a lab test: 45.3 seconds, 38.9 seconds, 35.6 seconds, and
41.1 seconds. What is the average number of seconds recorded?

a. 38.2 sec   **b.** 40.6 sec   **c.** 40.225 sec   **d.** 42.2 sec   **e.** 43 sec

Estimation will not help to solve this problem because all of the answer choices are very close together. Use a calculator as shown below.

Keys	Calculator displays
4 5 . 3	45.3
+	45.3
3 8 . 9	38.9
+	84.2
3 5 . 6	35.6
+	119.8
4 1 . 1	41.1
=	160.9
÷	160.9
4	4.
=	**40.225**

When you use a calculator during a test, keep the following in mind:

- Enter numbers and operations carefully; watch the display to be sure you have entered the numbers you want.

- ALWAYS check your calculations by repeating them, or by performing the opposite operation.

- Use your calculator to check *all* answers—even those you did not use a calculator for in the first place.

- Your thinking skills are as important as your calculator skills. A test is not necessarily easier just because you can use a calculator, so don't cut back on study time thinking that your calculator will guarantee success!

Using Mental Math

You have many tools and skills that help when you are taking a test. You can use a calculator for complex computation and use paper and pencil for less complex problems. For some problems, just using your head and your common sense will help. Before you do the problem below, ask yourself what makes sense.

EXAMPLE A nurse practitioner learned that her patient had lost 24 pounds since his last visit. She also noted that on that previous visit the patient had lost 18 pounds since his yearly physical. If the patient weighs 165 now, how much did he weigh at his yearly physical?

a. 118 lb   **b.** 123 lb   **c.** 183 lb   **d.** 189 lb   **e.** 207 lb

To solve this problem, you first have to use your thinking skills. If you focused on the word *lost* and quickly subtracted 24 and 18 from 165 with your calculator, you probably chose **b. 123 lb** as your answer. This choice is **incorrect.**

Think about the patient. Does he weigh more now or less? Since you are told he lost weight, he now weighs less. Therefore, to find his previous weight, it makes sense that you should *add* 24 and 18 to 165 to find his previous, heavier weight.

If you use your mental math skills, you'll quickly see that whatever the sum is, it is sure to be over 200. (Round 24 to 20 and 18 to 20. Add 20 + 20 = 40. Add 40 + 165 = 205.) Only **e. 207** could be correct.

Doing calculations in your head, instead of with pencil or a calculator, can save time on a test. However, mental math skills take awhile to develop. Get plenty of practice using this skill BEFORE you try using it in a test-taking situation. And always check your answers!

Using Estimation

Estimating is an excellent skill to use when you are taking a test. Here are three ways to use estimation to increase your chance of success on any math test.

1. **If a question uses words such as *about* or *approximately,* you are being told that you do not have to find an exact answer.**

   EXAMPLE   An overdue charge for a movie rental is $2.90 per day. A regular overnight rental is $3.00. About how much money will Mrs. Edwards pay if she rents two movies and pays for one movie that is 4 days overdue?

   a. $6   **b.** $8   **c.** $12   **d.** $16   **e.** $18

   SOLUTION   You do not have to multiply $2.90 by 4 to get the total overdue charge because the problem says "about how much money." Round $2.90 up to $3.00.

   $(4 \times \$3) + (2 \times \$3) = \$12 + \$6 = \$18$ **e. $18** is correct.

2. **Sometimes a good estimate will allow you to choose the correct answer.**

   EXAMPLE   A rectangular wall-hanging measures 35 inches across and 22 inches high. How many *feet* of trim are needed to go around the entire perimeter?

   a. 4.75 feet   **b.** 5.5 feet   **c.** 9.5 feet   **d.** 57 feet   **e.** 114 feet

   SOLUTION   One solution is to add the total inches of the four sides and divide by 12 to get the number of feet. A quicker way is to estimate.

   35 inches is about 3 feet. 22 inches is about 2 feet.
   $(2 \times 3) + (2 \times 2) = 10$ feet = approximate perimeter

   Choose the answer closest to your estimate of 10 feet. **c. 9.5 feet** is correct.

3. **Use an estimate to check an answer after you've done the computation.**

   EXAMPLE   A recent poll showed that 47% of all milk products purchased at the Health Stop Market were low-fat. If a total of $2,385 was spent on milk products in one week, how much money was spent on low-fat items?

   a. $298.40   **b.** $450.90   **c.** $665.35   **d.** $1,120.95   **e.** $1,994.20

   SOLUTION   Perform the calculations on your calculator ($.47 \times 2,385 = \$1,120.95$).

   Estimate to make sure that you have entered the numbers correctly. 47% is almost 50%, or half. $2385 \div 2 = 1,192.5$, which is close to the exact answer.

Formulas

AREA

Figure Name	Formula	Meaning
square	$A = s^2$	where s = side
rectangle	$A = lw$	where l = length, w = width
parallelogram	$A = bh$	where b = base, h = height
triangle	$A = \frac{1}{2}bh$	where b = base, h = height
circle	$A = \pi r^2$	where π = 3.14, r = radius

PERIMETER

square	$P = 4s$	where s = side
rectangle	$P = 2l + 2w$	where l = length, w = width
triangle	$P = a + b + c$	where a, b, and c are the sides
circumference (C) of a circle	$C = \pi d$	where π = 3.14, d = diameter

VOLUME

cube	$V = s^3$	where s = side
rectangular container	$V = lwh$	where l = length, w = width, h = height
cylinder	$V = \pi r^2 h$	where π = 3.14, r = radius, h = height
Pythagorean relationship	$c^2 = a^2 + b^2$	where c = hypotenuse, a and b are legs of a right triangle
simple interest (i)	$i = prt$	where p = principal, r = rate, t = time

EQUIVALENCIES

12 inches = 1 foot	16 ounces = 1 pound	60 seconds = 1 minute	8 ounces = 1 cup
3 feet = 1 yard	2,000 pounds = 1 ton	60 minutes = 1 hour	2 cups = 1 pint
5,280 feet = 1 mile	1 inch = 2.54 centimeters	24 hours = 1 day	2 pints = 1 quart
1,760 yard = 1 mile	1 quart = .95 liter	365 days = 1 year	4 quarts = 1 gallon

Glossary

A

area The square measure of a flat surface found by multiplying the length by the width

average The middle value of a group of numbers. An average is found by adding a group of numbers and dividing the total by the number of items in the group. To find the average of 35, 25, 48,

$$35 + 25 + 47 = 117 \quad \frac{117}{3} = 39$$

C

circumference Total distance around a circle

compare To decide whether one number is equal to, less than, or greater than another number. $3.7 > 3\frac{1}{2}$ means "three and seven tenths is greater than 3 and a half."

convert To change from one unit of measurement to another. To change 60 inches to feet,

$$\frac{60}{12} \text{ (inches per foot)} = 5 \text{ feet}$$

cross product The product of the numerator of one fraction and the denominator of another fraction. $\frac{2}{3} \times \frac{3}{4}$ 2×4 is one cross product; 3×3 is the other.

D

decimal A value smaller than 1, written to the right of a decimal point. In the number 12.125, .125 is a decimal and is smaller than 1.

denominator The bottom number in a fraction. It shows the number of parts a whole is divided into. In the fraction $\frac{3}{4}$, 4 is the denominator.

E

equivalent fractions Fractions that are equal in value. $\frac{1}{2}$ and $\frac{2}{4}$ are equivalent fractions.

estimate To find an approximate number, a number that is close to the actual value. An estimate for 597 + 101 would be 600 + 100, or 700.

equation A mathematical statement that two values are equal. $3 + 8 = 11$ is an equation. $27 - x = 20$ is an equation.

expression The use of math symbols to represent the relationship between numbers. The expression $\frac{x}{25}$ means x is being divided by 25.

F

fraction A number smaller than 1, written as part over whole. $\frac{3}{4}$ is a fraction (3 is the part; 4 is the whole).

friendly numbers Numbers that are easy to work with. 100 and 10 are easy numbers because one divides easily into the other.

function keys The buttons on a calculator that indicate computation operations, such as addition, subtraction, multiplication, and division.

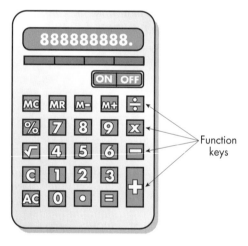

Function keys

I

improper fraction A fraction in which the numerator is the same size or greater than the denominator. $\frac{5}{5}$ and $\frac{8}{3}$ are improper fractions.

inverse operations Opposite operations; operations that undo each other. Subtraction and addition are inverse operations, as are multiplication and division.

L

lowest terms Not able to be reduced. A fraction is said to be in lowest terms when the numerator and denominator cannot be reduced further. $\frac{1}{5}$ is in lowest terms; $\frac{2}{10}$ is not.

M

mixed number A number that includes both a whole number and a fraction. $2\frac{1}{3}$ is a mixed number.

multiple choice A question or problem that offers two or more possible answers.

multistep problem A problem that takes more than one operation to solve.

N

numerator The top number in a fraction. The numerator tells you how many parts of a whole there are. In the fraction $\frac{2}{3}$, 2 is the numerator.

O

order of operations The acceptable order in which to perform math operations in a multistep problem. In the problem $\frac{(5+4)}{3}$, you must add the numbers in parentheses before you divide.

ordering The process of putting numbers in order from least to greatest or greatest to least.

P

percent A whole that is divided into one hundred equal parts. % = percent. 43% means 43 out of 100.

perimeter The total distance around a figure

place value The position a digit holds in a number. In 27,439, the 7 is in the thousands place value.

process A step-by-step method for completing a task

proper fraction A fraction in which the numerator is smaller than the denominator. $\frac{2}{3}$ is a proper fraction; $\frac{8}{5}$ is not.

proportion Two equal fractions or ratios. $\frac{3}{4} = \frac{6}{8}$ is a proportion.

R

ratio A comparison of one number to another. $\frac{4}{5}$, 4:5, and 4 to 5 are all ratios comparing 4 to 5.

remainder The part smaller than a whole that is left over in a division problem.

$$8\overline{)30} \quad \begin{array}{r} 3 \\ \underline{24} \\ 6 \text{ remainder} \end{array}$$

rounding Changing the value of a number slightly so that it ends in a number that is easy to work with. 89 can be rounded to 90; 2.03 can be rounded to 2.

S

set-up problem A problem that does not require computation. Instead, it requires you to choose the correct method for solving a problem.

V

variable A letter used to represent a number you do not know in a problem. $35 + y = 194$ (y is the variable)

W

word problem A math problem that asks a question in words

Index

A

Arithmetic expression 44, 131

C

Calculator 20, 137, 206
Charts 18, 75, 158, 166
Checking answers 34, 144, 151+
Comparing numbers 133
Conversions 133
Cross products 105

D

Division bar 48

E

Estimation 22, 121, 123, 127, 208
    with fractions 125
Equations 85, 89, 93, 97
Equivalencies 42, 209
Extra information 61

F

Five-step process 24
Formulas 173, 178, 209
Friendly numbers 123

G

Geometry 169+
Graphs 16, 156, 160, 163

H

Hidden information 57

I

Item sets 71, 75

L

Labels 55

M

Mental math 207
Missing information 69
Multiple-choice problems 12
Multistep problems 14

N

Not enough information 65

O

Operations 30, 82
Ordering numbers 133

P

Parentheses 47
Pictures 110, 156, 169, 180
Proportions 101, 105, 107

R

Ratios 101
Reasonable answers 148
Remainders 139
Rounding 22, 127, 140

S

Set-up problems 44, 47, 129

V

Variables 131